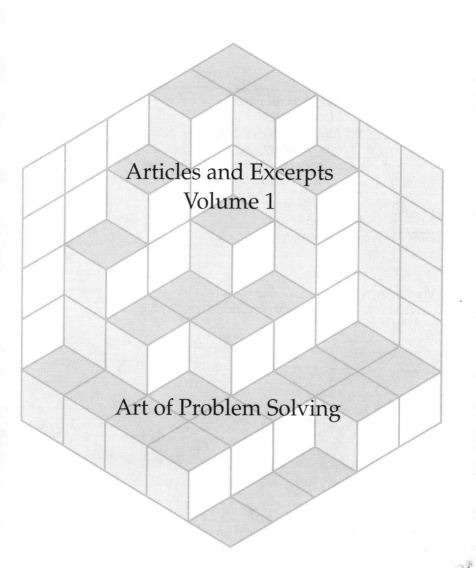

Articles and Excerpts
Volume 1

Art of Problem Solving

Published by: AoPS Incorporated
                  P.O. Box 2185
                  Alpine, CA 91903-2185
                  (619) 659-1612
                  books@artofproblemsolving.com

ISBN-10: 1-934124-00-1
ISBN-13: 978-1-934124-00-0

Visit the Art of Problem Solving website at
            http://www.artofproblemsolving.com

Prepared using the LaTeX document processing system.
Diagrams prepared using METAPOST.

Printed in the United States of America.

First Printing 2006.

Contents

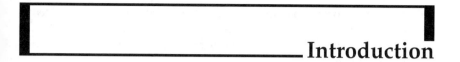

Introduction

What Is This Booklet?

This booklet contains chapters that are excerpted from the Art of Problem Solving's website and introductory subject textbooks. In 2005, the Art of Problem Solving began publishing a series of introductory textbooks:

- *Introduction to Counting & Probability* by David Patrick

- *Introduction to Geometry* by Richard Rusczyk

- *Introduction to Number Theory* by Mathew Crawford

In addition, the *Introduction to Algebra* textbook will be available in early 2007. Besides the material presented in this booklet, the full textbooks contain the following features:

➤ Each book has an accompanying Solution Manual, which contains **full solutions** to every Exercise, Review Problem, and Challenge Problem.

➤ Many problems in the books have one or more **hints** that can guide students who are stuck but who don't want to resort to just looking up the solution.

➤ The books are in **color**.

You can view sample pages of the textbooks, and purchase the books, on our website at

> http://www.artofproblemsolving.com

Learn By Solving Problems

We believe that the best way to learn mathematics is to solve problems. Lots and lots of problems. In fact, we believe that the best way to learn mathematics is to try to solve problems that you don't know how to do. When you discover something on your own, you'll learn it much better than if someone just tells it to you.

In the Art of Problem Solving textbooks, most of the sections of the book begin with several problems. The solutions to these problems will be covered in the text, but readers should try to solve the problems *before* reading the section. If you can't solve some of the problems, that's OK, because they will all be fully solved as you read the section. Even if you solve all of the problems, it's still important to read the section, both to make sure that your solution is correct, and also because you may find that the book's solution is simpler or easier to understand than your own.

If you find that the problems are too easy, this means that you should try harder problems. Nobody ever learns very much by solving problems that are too easy for them.

Resources

Here are some other good resources for you to further pursue your study of mathematics and problem solving:

- *The Art of Problem Solving* books, by Sandor Lehoczky and Richard Rusczyk. These books, now in their 7th edition, cover a wide range of problem solving topics across many different areas of mathematics.

- The www.artofproblemsolving.com website. The authors of this booklet are also the webmasters of the Art of Problem Solving website, which contains many resources for students: a discussion forum; online classes; a problem solver's Wiki; articles; resource lists of books, contests, and other websites; a LATEX tutorial; and much more!

- You can hone your problem solving skills (and perhaps win prizes!) by participating in various mathematical contests. For middle school students in the United States, two major contests are MATHCOUNTS and the AMC 8. For U.S. high school students, some of the best-known contests are the AMC/AIME/USAMO series of contests (which choose the U.S. team for the International Mathematics Olympiad) and the USA Mathematical Talent Search. More details about these and many other contests are available on the Art of Problem Solving website.

Contests

We would like to thank the following contests for allowing us to use a selection of their problems in this book:

- The American Mathematics Competitions (www.unl.edu/amc)

- MATHCOUNTS® (www.mathcounts.org)

- The Mandelbrot Competition (www.mandelbrot.org)

- The Harvard-MIT Mathematics Tournament (web.mit.edu/hmmt)

- The USA Mathematical Talent Search (www.usamts.org)

- The American Regions Math League (www.arml.com)

More information about all of these contests can be found on the contests' individual websites, and at the Art of Problem Solving website at www.artofproblemsolving.com.

Thanks

Thanks to Amanda Jones, Brian Rice, Naoki Sato, and Vanessa Rusczyk for significant help with the diagrams, layout, and proofreading.

CHAPTER 1

Three Types of Probability

Note: This article was written by Mathew Crawford. This article and other articles about problem solving can be found at www.artofproblemsolving.com in the "Resources" section.

This article is not so much about particular problems or problem solving tactics as it is about labels. If you think about it, labels are a big key to the way we organize ideas. When we already have the central concepts to problems organized, we are better able to solve them and our solutions are often more efficient. In short, labels help us organize – and organization simplifies problem solving! This article seeks to demonstrate the power of intelligent classification using types of probability as an example.

1.1 Introduction

Each day on his way to work, Steve drives up to the busy four-way intersection in Omaha. When the traffic light signals green, Steve drives through the intersection. When the traffic light is red, he stops and waits for it to turn green. When the traffic light is yellow, Steve considers whether or not he will make it through the intersection in time before making a decision as to whether to stop or go. It doesn't take a whole lot of effort for Steve to make it into work.

Without the traffic light in place, making it through the intersection might be a chore and it might not even be possible. He would always have to slow down, prepared to stop if necessary. He'd need to look around to see if there are cars coming from the other three directions that might cross his path. If there are enough other drivers, the whole process would be chaos!

It's a good thing we have traffic lights to make driving easier.

Now, let's build a probability traffic light! We can classify three main types of probability problems based on the ways in which we can approach them: **counting**, **geometry**, and **algebra**. When we can identify these types as easily as the colors on a traffic light, we can cut to the chase and solve problems.

1.2 Probability as Counting

The first type of probability we will discuss is perhaps the simplest to understand. Let P(event) be the probability of some event occuring. We can often determine P(event) by counting the number of successful outcomes and then dividing by the total number of equally likely outcomes:

> **Concept:**
> $$P(\text{event}) = \frac{\# \text{ of successful outcomes}}{\# \text{ of total outcomes}}$$

Let's take a look at a couple of problems that apply this principle of counting to solve probability problems.

> **Problem 1.1:** Find the probability that when two standard 6-sided dice are rolled, the sum of the numbers on the top faces is 5.

Solution for Problem 1.1: There are $6 \cdot 6 = 36$ possible outcomes when we roll a pair of dice. We can list the outcomes in which the sum of the top faces is 5:

$$\boxed{\cdot} + \boxed{\vcenter{\hbox{$\cdot\ \cdot$}}} = 5$$
$$\boxed{\vcenter{\hbox{$\cdot\ \cdot$}}} + \boxed{\vcenter{\hbox{$\cdot\ \cdot$}}} = 5$$
$$\boxed{\vcenter{\hbox{$\cdot\ \cdot$}}} + \boxed{\vcenter{\hbox{$\cdot\ \cdot$}}} = 5$$
$$\boxed{\vcenter{\hbox{$\cdot\ \cdot$}}} + \boxed{\cdot} = 5$$

We can now reach an answer by dividing the number of successful outcomes by the total number of possible outcomes:

$$P(\text{Sum of 5}) = \frac{\text{\# of successful outcomes}}{\text{\# of total outcomes}} = \frac{4}{36} = \frac{1}{9}.$$

□

Now let's take a look at another example of counting probability that requires a bit more thought:

Problem 1.2: A bag contains 16 marbles, 4 of which are blue and 12 of which are green. Two marbles are randomly pulled from the bag at the same time. What is the probability that both marbles are blue?

Solution for Problem 1.2: We can work this problem in several ways – all of which are based in counting methods.

In our first solution we note that it doesn't matter that both marbles are drawn at once. We can arbitrarily call one of them the *first marble* and the other the *second marble*. There are 16 choices for the first marble, leaving 15 choices for the second marble. The total number of ways in which we can draw two marbles from the bag is thus $16 \cdot 15 = 240$.

Now we count the number of ways in which we can draw two blue marbles from the bag in two draws. There are 4 choices for the first blue marble, leaving 3 for the second blue marble, for a total of $4 \cdot 3 = 12$ ways in which we can draw two blue marbles from the bag.

We now calculate the probability:

$$P(\text{2 blue marbles}) = \frac{\text{\# of ways to draw 2 blue marbles}}{\text{\# of ways to draw 2 marbles}} = \frac{12}{240} = \frac{1}{20}.$$

Students familiar with combinations might solve this problem in a related manner. If we don't consider the order in which we select the marbles, we can note that there are $\binom{16}{2}$ ways in which we can select two marbles and $\binom{4}{2}$ ways in which we can select two blue marbles. Now our calculation looks like this:

$$P(\text{2 blue marbles}) \quad = \quad \frac{\text{\# of ways to draw 2 blue marbles}}{\text{\# of ways to draw 2 marbles}}$$

$$= \quad \frac{\binom{4}{2}}{\binom{16}{2}} = \frac{\frac{4 \cdot 3}{2}}{\frac{16 \cdot 15}{2}} = \frac{6}{120} = \frac{1}{20}$$

We could even solve this problem by calculating the probability one draw at a time and multiplying. The probability that the first marble is blue is $\frac{4}{16} = \frac{1}{4}$. If the first marble was blue, then 3 of the remaining 15 marbles are blue, so the probability that the second marble will also be blue is $\frac{3}{15} = \frac{1}{5}$. The probability that both are blue is the product of the probabilities of each draw:

$$P(\text{2 blue marbles}) \quad = \quad P(\text{first marble is blue}) \cdot P(\text{second marble is blue})$$

$$= \quad \frac{1}{4} \cdot \frac{1}{5} = \frac{1}{20}$$

☐

Counting probability problems are fairly common and they are the easiest to recognize. When we can count successful and total numbers of equally likely outcomes, the probability is the ratio between the two.

1.3   Probability as Geometry

Sometimes it is impossible to count the total number of events because there are infinitely many possibilities. Some problems involve evaluating points on a line segment or in space, or particular values of a continuous variable such as time. The following is a simple example:

> **Problem 1.3:**   Lawrence parked his car in a parking lot at a randomly chosen time between 2:30 PM and 4:00 PM. Exactly half an hour later he drove his car out of the parking lot. What is the probability that he left the parking lot after 4:00 PM?

Solution for Problem 1.3:   We can divide hours into minutes, minutes into seconds, and seconds into smaller and smaller fractions. There are an infinite number of times during which Lawrence could have driven away. We cannot count times to solve this problem!

Time is continuous, so we'll have to think differently about this problem. Let's take a look at a line segment that represents the possible times Lawrence could drive away:

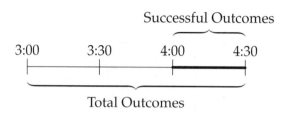

We can determine the probability that Lawrence drove away after 4:00 PM by comparing the length of the segment of successful outcomes to that of the total outcomes:

$$P(\text{success}) = \frac{\text{length of successful segment}}{\text{length of total segment}} = \frac{30 \text{ (minutes)}}{90 \text{ (minutes)}} = \frac{1}{3}.$$

□

Now that we've seen one example of geometric probability, we can generalize the central idea:

> **Concept:** When a probability calculation involves one or more continuous variables, we can determine the probability by comparing the sizes of geometric regions:
>
> $$P(\text{event}) = \frac{\text{size of successful region}}{\text{size of total region}}.$$

Sometimes the sizes we compare are lengths, sometimes they are areas, and sometimes they are volumes.

Let's take a look at a problem involving multiple continuous variables:

Problem 1.4: If $1 \leq x \leq 4$ and $2 \leq y \leq 6$, find the probability that $x + y \geq 5$.

Solution for Problem 1.4: The problem is a little more complex because we have two distinct continuous variables. However, the concept is the same: we need to compare the successful region to the total region in order to determine the probability.

Let's examine the graph of the total region in which the successful region is shaded:

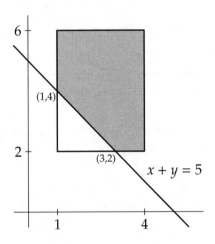

The regions we are comparing are areas. We determine the probability by comparing these areas:

$$\frac{\text{area of successful region}}{\text{area of total region}} = \frac{10}{12} = \frac{5}{6}.$$

Some adept problem solvers might notice that it's easier to compute the area of the unsuccessful region. We can calculate that probability and subtract it from 1:

$$P(\text{successful outcome}) = 1 - P(\text{unsuccessful outcome}) = 1 - \frac{2}{12} = \frac{5}{6}.$$

We call this technique **complementary probability** and it comes in handy quite often. □

1.4 Probability as Algebra

While it is rarely (if ever) taught in regular textbooks, we can solve some probability problems using algebraic techniques. The idea behind the next couple of problems is to view probabilities as variables. We then apply algebraic methods to find the values of those variables.

Let's take a look at a fairly simple example:

> **Problem 1.5:** If the probability that it rains next Tuesday in Seattle is twice the probability that it doesn't, what is the probability that it rains next Tuesday in Seattle?

Solution for Problem 1.5: We can neither count the rain nor represent it as a continuous variable. However, we can name the probability that it rains next Tuesday. That is, we can assign its value a variable.

Let x be the probability that it rains next Tuesday. We can now translate this word problem into a math problem in terms of x. Since it either will rain or won't rain next Tuesday in Seattle, the probability that it won't must be $1 - x$. We are told that
$$x = 2(1 - x).$$
Solving this equation we find that $x = \frac{2}{3}$, which is our answer. $\square$

> **Concept:** If you can identify enough relationships between probabilities, you might be able to solve the problem algebraically.

Let's take a look at a problem that requires a little more insight into algebraic relationships:

> **Problem 1.6:** Johnny and Michael play a game in which they take turns rolling a pair of fair dice until one of them rolls "snake eyes" (both dice show 1's). That person is the winner. If Johnny goes first, what is the probability that Johnny wins the game?

Solution for Problem 1.6: Let's go ahead and name the values of the probabilities that Johnny wins and that Michael wins. Let j be the probability that Johnny wins and m be the probability that Michael wins.

Since the probability that one of the two wins is 1 (make sure you see why the probability that the game is never won is 0), we have the equation $j + m = 1$. Unfortunately, this alone is not enough to solve the problem.

It often helps to explore some simple cases of a problem when you don't immediately see a path toward the solution. By examining the probabilities that Johnny and Michael each win during their first toss, second toss, etc., we notice that each of their probabilities is the sum of a geometric series:

$$
\begin{array}{ccccccccc}
& & \substack{\text{winner} \\ \text{on toss 1}} & & \substack{\text{winner} \\ \text{on toss 2}} & & \substack{\text{winner} \\ \text{on toss 3}} & & \cdots \\
j & = & \frac{1}{36} & + & \frac{1}{36}\left(\frac{35}{36}\right)^2 & + & \frac{1}{36}\left(\frac{35}{36}\right)^4 & + & \cdots \\
m & = & \frac{1}{36}\left(\frac{35}{36}\right) & + & \frac{1}{36}\left(\frac{35}{36}\right)^3 & + & \frac{1}{36}\left(\frac{35}{36}\right)^5 & + & \cdots
\end{array}
$$

While we could now find the value of j by summing a geometric series, a new relationship has presented itself! Notice that the ratio between the probability that Johnny wins and the probability that Michael wins is constant for every toss. This means that m is $\frac{35}{36} \cdot j$. We now have a system of linear equations in two variables:

$$
\begin{aligned}
j + m &= 1 \\
m &= \frac{35}{36} \cdot j
\end{aligned}
$$

We can solve this system by ordinary means to find that $j = \frac{36}{71}$. $\square$

1.5 Summary

Probability can get much harder than the problems we explored. However, our categorization of the areas of math that we can use to tackle these problems helps demystify probability at any level. Once we recognize these problems for what they are – counting, geometry, and algebra – we can be more confident and straight-forward in choosing the ways in which we approach probability problems.

Now that you see the power of intelligent classification, you can try to apply it to other areas of mathematics and other problem solving subjects!

1.6 Some Probability Practice Problems

Throughout the following probability problems, try first to determine the nature of the problem – whether it is a problem that can be solved using counting techniques, geometric techniques, algebraic techniques, or possibly a combination of more than one. Then try to solve each problem.

Exercises for Section 1.6

1.6.1 Olga and Andrew play chess each day. The probability that Olga beats Andrew on any given day is twice the probability that Andrew beats Olga. The probability that Andrew beats Olga is three times the probability that the game ends in a draw (tie). What is the probability that any particular game between Olga and Andrew ends in a draw?

1.6.2 A fair coin is flipped three times. What is the probability that at least one lands heads and at least one lands tails?

1.6.3 Of the 12 marbles in a bag, exactly 3 are blue. If Sandor reaches into the bag and pulls out two marbles, what is the probability that exactly one of them is blue?

1.6.4 Two real numbers x and y are randomly selected on a number line between 0 to 12. Find the probability that their sum is greater than 15.

1.6.5 The surface of an $8 \times 10 \times 12$ block is painted green after which the block is cut up into 960 smaller $1 \times 1 \times 1$ cubes. If one of the smaller cubes is selected at random, what is the probability that it has green paint on at least one of its faces?

1.6.6 Ted and Erin play a game in which they take turns flipping a fair coin. The first player to flip heads wins. If Erin goes first, what is the probability she wins?

1.6.7 Bender orders two robot painters to paint a 1000 meter long circular fence. One robot painter is ordered to paint a randomly selected 720 meter stretch of the fence while the other is ordered to paint a randomly selected 750 meter stretch of the fence. Find the probability that when the robot painters are done, the entire fence has been painted.

1.6.8 A point (x, y) in the coordinate plane satisfies the inequalities

$-6 < x < 6$ and $-6 < y < 6$, where x and y are real numbers. Find the probability that the point (x, y) is less than 4 units from the origin.

1.6.9★ A lattice point (x, y) is selected at random from the interior of the circle described by the equation $x^2 + y^2 - 2x + 4y = 22$. Find the probability that the point is within 1.5 units from the origin.

1.6.10★ An 'unfair' coin has a 2/3 probability of turning up heads. If this coin is tossed 50 times, what is the probability that the total number of heads is even? *(Source: AHSME)*

1.6.11★ Two mathematicians take a morning coffee break each day. They arrive at the cafeteria independently, at random times between 9 AM and 10 AM, and stay for exactly m minutes. Find the value of m if the probability that either one arrives while the other is in the cafeteria is 40%.
(Source: AIME)

1.6.12★ A fair coin is to be tossed ten times. Find the probability that heads never occur on consecutive tosses. *(Source: AIME)*

CHAPTER 2

Geometric Probability

Note: This chapter is excerpted from the book Introduction to Counting & Probability *by David Patrick.*

2.1 Introduction

Sometimes we are faced with probability problems in which there is no readily available quantity for us to count. One common example of this is in geometry. In geometric probability problems, instead of counting individual items or outcomes, we will use as our "counts" various geometric quantities, such as lengths or areas. These problems have a very different flavor to the problems that we've seen in this book until now, but the general technique is similar.

The basic idea will be to represent the set of all possibilities as some geometric object, and the set of successful possibilities as some portion of our first object. Then, we can calculate the probability of a successful outcome as

$$P(\text{success}) = \frac{\text{Size of successful region}}{\text{Size of possible region}},$$

where here, we've replaced the usual "number of items" by the concept of "size", which, depending on the problem, might be length, or area, or volume.

This idea may seem vague right now, but hopefully by the end of this chapter you'll better appreciate it. Geometric probability also gives us another important probability tool: we can often represent a non-geometric problem in terms of geometry, which we can then use to calculate probability.

Once again, the basic concept of geometric probability is:

Concept:

$$P(\text{success}) = \frac{\text{Size of successful region}}{\text{Size of total region}}.$$

2.2 Probability Using Lengths

Here's a basic problem to illustrate how we compute probability with lengths.

Problem 2.1: In Figure 2.1, $\overline{AC}$ has length 5, and $\overline{AB}$ has length 4. A point P is selected randomly on the segment $\overline{AC}$. What is the probability that P is closer to B than to A?

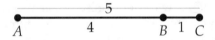

Figure 2.1: Line segment for Problem 2.1

Solution for Problem 2.1: We first need to determine the region in which the condition is satisfied – in other words, in what region are the points closer to B than to A. If we let D be the midpoint of the segment $\overline{AB}$, then it is clear that all of the points along $\overline{DC}$, except D itself, are closer to B than to A, and conversely all of the points along $\overline{AD}$, except D itself, are closer to A than to B. (D is equidistant from A and B.) We can see this visually in Figure 2.2.

Figure 2.2: Regions of the line segment

So the "successful region" is the segment $\overline{DC}$ (except for D), and the "possible region" is the segment $\overline{AC}$. Therefore:

$$P(P \text{ closer to } B \text{ than to } A) = \frac{\text{Length of successful region}}{\text{Length of possible region}}$$
$$= \frac{\text{Length of } \overline{DC}}{\text{Length of } \overline{AC}}$$
$$= \frac{3}{5}.$$

□

One important thing to note regarding the solution to Problem 2.1 is that when measuring the successful region, the point D by itself doesn't contribute anything to the length. So it's OK to use the length of $\overline{DC}$ as the size of our successful region, because the length of the segment $\overline{DC}$ including D is the same as the length of $\overline{DC}$ without the point D.

The next problem is an example of how we can use geometry in problems that don't at first appear geometric.

Problem 2.2: A real number x is selected randomly such that $0 \le x \le 3$. What is the probability that $|x - 1| \le \frac{1}{2}$?

Solution for Problem 2.2: We can easily see that for x to satisfy the required condition, we must have $\frac{1}{2} \le x \le \frac{3}{2}$. So the "successful region" is the segment of the real line that runs from $\frac{1}{2}$ to $\frac{3}{2}$. This region has length 1. The "possible region" is the segment of the real line that runs from 0 to 3, and has length 3. Therefore,

$$P\left(|x - 1| \le \frac{1}{2}\right) = \frac{\text{Length of segment with } \frac{1}{2} \le x \le \frac{3}{2}}{\text{Length of segment with } 0 \le x \le 3} = \frac{1}{3}.$$

□

Notice that in these two problems (Problems 2.1 and 2.2), we couldn't count outcomes, because our successful region and total region were **continuous**, not **discrete**. (**Discrete** basically means you can count them 1,2,3,....) However, we can still use our principles of probability because we can measure the size of the regions through length.

> **Concept:** If we are trying to compute a probability in which the outcomes are continuous (as opposed to discrete), then we use geometric probability.

> **Problem 2.3:** A real number x is chosen at random such that $0 < x < 100$. What is the probability that $\lfloor \sqrt{x} \rfloor$ is even? (Note: for any real number y, $\lfloor y \rfloor$ is defined to be the greatest integer less than or equal to y.)

Solution for Problem 2.3: The "total possible" region is just $0 < x < 100$, which is a segment of length 100.

To compute the regions corresponding to "successful" outcomes, we can set up a table:

$\lfloor \sqrt{x} \rfloor$	Interval	Length
0	$0 \le x < 1$	1
1	$1 \le x < 4$	3
2	$4 \le x < 9$	5
3	$9 \le x < 16$	7
4	$16 \le x < 25$	9
5	$25 \le x < 36$	11
6	$36 \le x < 49$	13
7	$49 \le x < 64$	15
8	$64 \le x < 81$	17
9	$81 \le x < 100$	19

We can see that the total of the lengths of the intervals with the successful outcomes is $1 + 5 + 9 + 13 + 17 = 45$, so the probability is $\frac{45}{100} = \frac{9}{20}$. $\square$

Exercises for Section 2.2

2.2.1 Let $\overline{AB}$ be a line segment of length 10. A point P is chosen at random on $\overline{AB}$. What is the probability that P is closer to the midpoint of $\overline{AB}$ than to either endpoint?

2.2.2 Let $\overline{CD}$ be a line segment of length 6. A point P is chosen at random on $\overline{CD}$. What is the probability that the distance from P to C is smaller than the *square* of the distance from P to D?

2.2.3 A real number x is chosen at random such that $-2 \leq x \leq 5$. What is the probability that $x^2 < 2$?

2.2.4 A real number y is chosen at random such that $0 \leq y \leq 100$. What is the probability that $y - \lfloor y \rfloor \geq \frac{1}{3}$?

2.2.5★ A real number z is chosen at random such that $0 \leq z \leq 1$. What is the probability that, when written as a (possibly infinite) decimal in base 10, that the second digit to the right of the decimal point is a prime number?

2.2.6★ Let P be a point chosen at random on the line segment between the points $(0, 1)$ and $(3, 4)$ on the coordinate plane. What is the probability that the area of the triangle with vertices $(0, 0)$, $(3, 0)$ and P is greater than 2?

2.3 Probability Using Areas

We can calculate probabilities using areas in pretty much the same way that we calculate probabilities using lengths. As with lengths, we will have a region corresponding to "total outcomes." Within this region, we will have a smaller region corresponding to "successful outcomes." Then

$$P(\text{success}) = \frac{P(\text{Area of successful outcomes region})}{P(\text{Area of total outcomes region})}.$$

Areas are in fact a much more powerful tool than mere lengths, and we can model a great many more problems using areas. Let's look at a few examples.

> **Problem 2.4:** Point C is chosen at random atop a 5 foot by 5 foot square table. A circular disk with a radius of 1 foot is placed on the table with its center directly on point C. What is the probability that the entire disk is on top of the table (i.e. that none of the disk hangs over an edge of the table)?

Solution for Problem 2.4: The "total outcomes" region is easy: it's just the surface of the table, so it's a square region with side length 5.

The "successful outcomes" region is trickier. We can draw a diagram as in Figure 2.3.

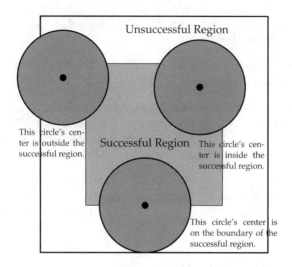

Figure 2.3: Table and some disks for Problem 2.4

We can see that the disk will be entirely on the table if and only if C is at least 1 foot away from each edge of the table. Therefore, C must be within a central square region of side length 3, as shown in Figure 2.3.

So now we can compute the probability:

$$P(\text{success}) = \frac{P(\text{Area of successful outcomes region})}{P(\text{Area of total outcomes region})} = \frac{3 \times 3}{5 \times 5} = \frac{9}{25}.$$

$\square$

Problem 2.4 is an example of a problem which is stated in terms of geometry. But we can also use areas to solve algebraic probability problems similar to Problem 2.2.

> **Problem 2.5:** Suppose two numbers x and y are each chosen such that $0 < x < 1$ and $0 < y < 1$. What is the probability that $x + y > \frac{3}{2}$?

Solution for Problem 2.5: Since there's nothing discrete in this problem that can be counted, we'll need to use geometry.

The "total outcomes" region is the region in the xy-plane with $0 < x < 1$ and $0 < y < 1$. This is the interior of the square with corners $(0,0)$, $(0,1)$, $(1,1)$, and $(1,0)$, as shown in Figure 2.4.

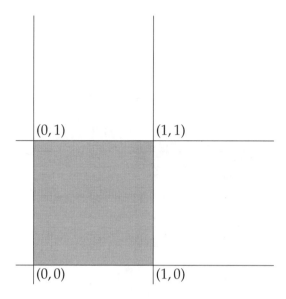

Figure 2.4: Total outcomes region for Problem 2.5

Now we need to describe the region corresponding to successful outcomes. We need $x + y > \frac{3}{2}$, so this will be the region inside the square of Figure 2.4 which is above the line $x + y = \frac{3}{2}$. We will add this to our diagram in Figure 2.5, and see that our successful region is the interior of a triangle.

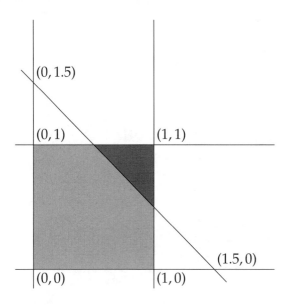

Figure 2.5: Total and successful outcomes regions for Problem 2.5

We can now calculate the areas. The area of the square is 1. The area of the triangle is $\frac{1}{8}$. Therefore the probability is

$$P(\text{success}) = \frac{P(\text{Area of successful outcomes region})}{P(\text{Area of total outcomes region})} = \frac{\frac{1}{8}}{1} = \frac{1}{8}.$$

□

The real power of geometric methods is taking word problems and recasting them as geometry problems that we can work with. The next problem is a classic example of this.

Problem 2.6: My friend and I are hoping to meet for lunch. We will each arrive at our favorite restaurant at a random time between noon and 1 p.m., stay for 15 minutes, then leave. What is the probability that we will meet each other while at the restaurant? (For example, if I show up at 12:10 and my friend shows up at 12:15, then we'll meet; on the other hand, if I show up at 12:50 and my friend shows up at 12:20, then we'll miss each other.)

Solution for Problem 2.6: Once again, because we have infinitely many possibilities, we won't be able to use regular counting, so we'll need to look for a geometric approach.

We can make a graph plotting my arrival time and my friends arrival time. We'll put my arrival time on the *x*-axis and my friend's arrival time on the *y*-axis. The result is Figure 2.6.

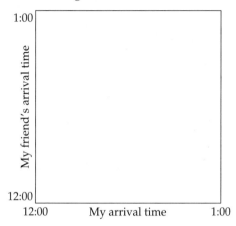

Figure 2.6: Graph for meeting times

We now need to figure out what portion of Figure 2.6 corresponds to me and my friend meeting. If my friend and I are to successfully meet, I must arrive no earlier than 15 minutes before he arrives, and no later than 15 minutes after he arrives. If unsure how to proceed at this point, we could experiment with a few example arrival times to see what the picture is going to look like.

Concept: If unsure how to proceed with a problem, try a few simple examples.

Here are a few examples:

My arrival time	My friend's successful arrival times
12:00	12:00-12:15
12:10	12:00-12:25
12:20	12:05-12:35
12:30	12:15-12:45
12:40	12:25-12:55
12:50	12:35-1:00
1:00	12:45-1:00

We can plot these on our graph, and the result is shown in Figure 2.7.

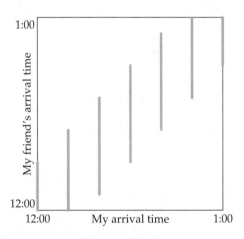

Figure 2.7: Some possible meeting times

Based on what we've drawn in Figure 2.7, we have a pretty good picture of what the successful meeting region is going to look like. We can essentially "fill in" the picture, to get Figure 2.8.

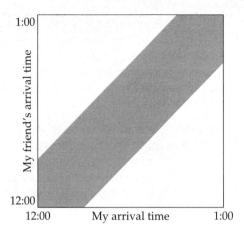

Figure 2.8: Successful meetings

To be precise, we can write the condition for a successful meeting as an equation:
$$y - 15 \leq x \leq y + 15.$$

So the "successful outcomes" region is the region inside the square that's below the line $y = x + 15$ and above the line $y = x - 15$. Figure 2.9 shows the result after we add these lines to our diagram.

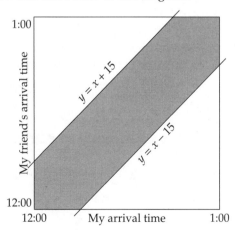

Figure 2.9: Successful meeting times

Figure 2.9 shows a two-dimensional representation of all possible times my friend and I can show up, along with a shaded band representing success. Notice, as expected, that the graph is symmetrical. Because there is nothing in the problem to indicate a difference between my situation and that of my friend, we should expect to see no difference if we were to swap axes.

We can now compute the areas of the two regions, and then divide to get the probability. The "total outcomes" region is just a square with side length 60, so its area is $60 \times 60 = 3600$. The "successful outcomes" region is a hexagon, for which it is somewhat unpleasant to calculate the area. But we can pretty easily calculate the area of the "unsuccessful outcomes" region, which is just two triangles. Each half of the region is a right isosceles triangle with side length 45, with area $\frac{1}{2} \times 45 \times 45$. So the "unsuccessful outcomes" region has total area $2 \times \frac{1}{2} \times 45 \times 45 = 2025$.

Therefore the "successful outcomes" region has area $3600 - 2025 = 1575$, and finally the probability of success is

$$P(\text{we meet for lunch}) = \frac{1575}{3600} = \frac{7}{16}.$$

□

Exercises for Section 2.3

2.3.1 A point (x, y) is chosen at random inside the square with vertices $(0, 0)$, $(0, 1)$, $(1, 1)$, and $(1, 0)$. What is the probability that

(a) $x + y \leq 0.5$?

(b) $x + 2y \geq 1$?

(c) $|x - y| \leq 0.2$?

(d) $x^2 + y^2 < 1$?

(e) the distance from (x, y) to the center $(0.5, 0.5)$ of the square is less than 0.5?

(f) the distance from (x, y) to $(0, 1)$ is greater than 1?

2.3.2 Steve's kitchen floor has a tile pattern of square tiles of side length 10 cm. Steve drops a penny (which has radius 1 cm) on the floor. What is the probability that the penny lies entirely within one tile?

2.3.3 Repeat Problem 2.6, except this time my friend will wait for 20 minutes (while I will still only wait for 15 minutes).

2.3.4 Maryanne's mail arrives at a random time between 1 p.m. and 3 p.m. Maryanne chooses a random time between 2 p.m. and 3 p.m. to go check her mail. What is the probability that Maryanne's mail has been delivered when she goes to check on it?

2.4 Summary

➤ We can use geometric tools to solve certain probability problems. The basic formula is

$$P(\text{success}) = \frac{\text{Size of successful region}}{\text{Size of total region}}.$$

➤ Geometric probability can be used in situations where we don't have a set of individual items to "count"; rather, we have infinitely many outcomes which can be nicely represented using some geometric object (such as a line segment, a region of a plane, etc.). A more technical way of saying this is that we use geometric probability when our outcomes are continuous rather than discrete.

Some useful ideas:

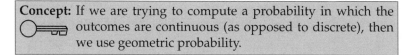

Concept: If we are trying to compute a probability in which the outcomes are continuous (as opposed to discrete), then we use geometric probability.

Concept: If unsure how to proceed with a problem, try a few simple examples.

Review Problems for Chapter 2

2.7 Let x be a real number randomly chosen so that $-1 \leq x \leq 1$. What is the probability that $x^2 > \frac{1}{2}$?

2.8 Two points P and Q are randomly chosen on a circle C. What is the probability that the smaller arc between P and Q measures less than $60°$?

2.9 A point R is randomly chosen on the line segment from $(0,0)$ to $(5,10)$. What is the probability that the y-coordinate of R is at least 7?

2.10 A regular octagon $ABCDEFGH$ is given, and a point S on the octagon is chosen. What is the probability that S is closer to A than to any other vertex?

2.11 Two numbers x and y are chosen such that $0 \le x \le 1$ and $0 \le y \le 2$. What is the probability that:

  (a) $x \le y$?
  (b) $x + 1 \le y$?
  (c) $y^2 + x^2 > 1$?
  (d) $\frac{y}{x} > 5$?
  (e) the distance from the point $(x, y - 1)$ to the origin is less than 1?

2.12 A point is chosen at random inside the square with vertices $(0,0)$, $(2,0)$, $(2,2)$, and $(0,2)$. What is the probability that this point is closer to $(0,0)$ than to $(3,3)$?

2.13 Frankie rides the subway to work. She either takes the local or the express, whichever comes first. The local arrives every 15 minutes, at :00, :15, :30, and :45 past the hour. The express arrives every 20 minutes, at :05, :25, and :45 past the hours. (If both trains arrive at the same time, she takes the express.) Assuming that she arrives at the subway station at a random time, what is the probability that Frankie rides the express to work?

2.14 Given that a and b are real numbers such that $-3 \le a \le 1$ and $-2 \le b \le 4$, and values for a and b are chosen at random, what is the probability that the product ab is positive? *(Source: MATHCOUNTS)*

Challenge Problems for Chapter 2

2.15 Three points x, y, z are chosen at random on the unit interval $(0, 1)$. What is the probability that $x \le y \le z$?

2.16 A line segment is broken at two random points along its length.

What is the probability that the three new segments form a triangle? ,

2.17   A point P is randomly chosen in the interior of the right triangle ABC, as shown in Figure 2.10. What is the probability that the area of PBC is less than half of the area of ABC?

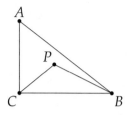

Figure 2.10: Figure for Problem 2.17

2.18★   Triangle ABC is a 30-60-90 right triangle with right angle at C, $\angle ABC = 60°$, and hypotenuse of length 2. Let P be a point chosen at random inside ABC, and extend ray BP to hit side AC at D. What is the probability that $BD < \sqrt{2}$?

2.19   Point P is chosen at random inside square $ABCD$.

(a)   What is the probability that triangle ABP has a greater area than triangle CDP?

(b)   What is the probability that triangle ABP has a greater area than each of triangles BCP, CDP, and DAP?

(c)★   What is the probability that ABP has a greater area than each of triangles BCP and CDP?

2.20★   What is the probability that if three points are chosen at random on the circumference of a circle, then the triangle formed by connecting the three points does not have a side with length greater than the radius of the circle? *(Source: AMC)*

2.21★   Three points are selected randomly on the circumference of a circle. What is the probability that the triangle formed by these three points contains the center of the circle?

USA
MATHEMATICAL
TALENT SEARCH

The USA Mathematical Talent Search is a **FREE** math competition open to middle and high school students in the United States.

The USAMTS consists of 4 rounds of 5 problems. Students have approximately *a month to solve each round of problems*, and may do research to solve the problems. Students must present full solutions, not just the answer.

> THE USAMTS IS AN OPPORTUNITY
> FOR STUDENTS TO
> *DEVELOP THEIR PROBLEM SOLVING
> SKILLS, IMPROVE THEIR TECHNICAL
> WRITING ABILITIES* AND *MATURE
> MATHEMATICALLY.*

In addition to book and software prizes, students can qualify for the American Invitational Math Exam (AIME) through the USAMTS.

Visit www.usamts.org for information about how to participate, past problems, submission information and more!

CHAPTER 3

Expected Value

Note: This article was written for the USA Mathematical Talent Search by David Patrick and Richard Rusczyk, and is reprinted with permission. Some portions of this article are reprinted from the book Introduction to Counting & Probability *by David Patrick. You can see the article in its original form at* www.usamts.org.

3.1 Introduction

You are probably already familiar with the concept of *mean* or *average*. For example, given the three numbers 5, 8, and 14, their average is $(5+8+14)/3 = 27/3 = 9$. More generally, if we are given numbers $x_1, x_2, \ldots, x_n$, then their average, often denoted $\bar{x}$, is:

$$\bar{x} = \frac{x_1 + x_2 + \cdots + x_n}{n}.$$

If we extend the concept of average to the outcome of a random event, we get what is called *expected value* or *expectation*. Expected value is the term we use to indicate the average result we would expect to get if we did a large number of trials for any experiment. Essentially, expected value is a weighted average, in which the more likely the outcome, the more that outcome is weighted.

Let's look at a simple example. Suppose that we are flipping a coin. If we get heads, we win $2, but if we get tails, we lose $1. If we flip the coin 1,000 times, we would expect to get heads 500 times and tails 500 times. We would win a total of $1,000 from our 500 heads and lose a total of $500 from our 500 tails, for a total net profit of $1,000 − $500 = $500. Since this profit is realized over the course of 1,000 flips, our average profit per flip is

$$\frac{\$500}{1000} = \$0.50.$$

We say that the expected value of each flip is $0.50. In this case, since heads and tails are equally likely, the expected value is just the usual average of the two outcomes:

$$\frac{+\$2 + (-\$1)}{2} = \frac{\$1}{2} = \$0.50.$$

However, if we are calculating the expected value of an event in which different outcomes occur with different probability, then we have to take a weighted average of the outcomes, as we will see a bit later.

3.2 Definition

We'll start by stating the formal definition, but don't be alarmed – the idea of expected value is not nearly as complicated as it may first appear.

Suppose that we have an event in which every outcome corresponds to some value. (An example is rolling a die: the "value" is the number showing on the top face of the die.) We have a list of possible values: $x_1, x_2, \ldots, x_n$. Value x_1 occurs with probability p_1, value x_2 occurs with probability p_2, and so on. Note that

$$p_1 + p_2 + \cdots + p_n = 1,$$

since the probabilities must total to 1. Then the *expected value* of the outcome is defined as the sum of the products of the outcomes' values and their respective probabilities:

$$E = p_1 x_1 + p_2 x_2 + \cdots + p_n x_n.$$

Traditionally, expected value is denoted by the capital letter E or by the Greek letter μ (called "mu").

Let's go back to our coin-flipping example, in which we win $2 for heads but lose $1 for tails. Since the probability of heads is $\frac{1}{2}$ and the probability of tails is also $\frac{1}{2}$, we can use the above formula to get:

$$E = \frac{1}{2}(\$2) + \frac{1}{2}(-\$1) = \$1 - \$0.50 = \$0.50,$$

which is what we got before.

3.3 Some simple examples

Now we'll do some fairly straightforward expected value computations. In all of these computations, we are simply taking the weighted average of the possible outcomes, in which the value of each outcome is weighted by the probability of that outcome.

Problem 3.1: What is the expected value of the roll of a standard 6-sided die?

Solution for Problem 3.1: Each outcome of rolling a 6-sided die has probability $\frac{1}{6}$, and the possible outcomes are ⚀, ⚁, ⚂, ⚃, ⚄, and ⚅. So the expected value is

$$\frac{1}{6}(1) + \frac{1}{6}(2) + \frac{1}{6}(3) + \frac{1}{6}(4) + \frac{1}{6}(5) + \frac{1}{6}(6) = \frac{21}{6} = 3.5.$$

Note that since each outcome in Problem 3.1 is equally likely, we can get the expected value simply by averaging the possible outcomes:

$$\frac{1 + 2 + 3 + 4 + 5 + 6}{6} = \frac{21}{6} = 3.5.$$

Also note that we can't actually roll a 3.5 on any individual roll. Thus, the expected value is not necessarily the most likely value, but rather the probability-weighted average of all possible values. □

Problem 3.2: Suppose you have a weighted coin in which heads comes up with probability $\frac{3}{4}$ and tails with probability $\frac{1}{4}$. If you flip heads, you win $2, but if you flip tails, you lose $1. What is the expected value of a coin flip?

Solution for Problem 3.2: By definition, we multiply the outcomes by their respective probabilities, and add them up:

$$E = \frac{3}{4}(+\$2) + \frac{1}{4}(-\$1) = \$1.50 - \$0.25 = \$1.25.$$

Another way to think of the expected value in Problem 3.2 is to imagine flipping the coin 1,000 times. Based on the probabilities, we would expect to flip heads 750 times and to flip tails 250 times. We would then win $1,500 from our heads but lose $250 from our tails, for a net profit of $1,250. Since this occurs over the course of 1,000 flips, our average profit per flip is

$$\frac{\$1,250}{1,000} = \$1.25.$$

□

Problem 3.3: In an urn, I have 20 marbles: 2 red, 3 yellow, 4 blue, 5 green, and 6 black. I select one marble at random from the urn, and I win money based on the following chart:

Color	Red	Yellow	Blue	Green	Black
Amount won	$10	$5	$2	$1	$0

What is the expected value of my winnings?

Solution for Problem 3.3: I draw a red marble with probability $\frac{2}{20}$, a yellow marble with probability $\frac{3}{20}$, and so on. Therefore, the expected value is

$$\frac{2}{20}(\$10) + \frac{3}{20}(\$5) + \frac{4}{20}(\$2) + \frac{5}{20}(\$1) + \frac{6}{20}(\$0) = \frac{\$48}{20} = \$2.40.$$

□

One common use of the expected value in a problem like Problem 3.3 is to determine a *fair price* to play the game. The fair price to play is the expected winnings, which is $2.40. In the long run, if I were charging people the fair price to play the game, I would expect to break even, neither making nor losing money. If I were running this game at a carnival, I could charge carnival-goers $2.50 to play the game, and I would expect to make a 10-cent profit, on average, from each person who plays.

Problem 3.4: Suppose I have a bag with 12 slips of paper in it. Some of the slips have a 2 on them, and the rest have a 7 on them. If the expected value of the number shown on a slip randomly drawn from the bag is 3.25, then how many slips have a 2?

Solution for Problem 3.4: We let x denote the number of slips with a 2 written on them. (This is the usual tactic of letting a variable denote what

---

we're trying to solve for in the problem.) Then there are $12 - x$ slips with a 7 on them.

The probability of drawing a 2 is $\frac{x}{12}$ and the probability of drawing a 7 is $\frac{12-x}{12}$, so the expected value of the number drawn is

$$E = \frac{x}{12}(2) + \frac{12-x}{12}(7) = \frac{84-5x}{12}.$$

But we are given that $E = 3.25$, so we have an equation we can solve for x:

$$3.25 = \frac{84-5x}{12}.$$

Solving this equation gives $x = 9$. Thus 9 of the 12 slips have a 2 written on them. □

3.4 Some more challenging problems

Now that we've finished describing expected value, we'll explore some more interesting problems. We'll solve a more challenging expected value problem, then use expected value as a tool to solve other problems.

> **Problem 3.5:** Suppose that 7 boys and 13 girls line up in a row. Let S be the number of places in the row where a boy and a girl are standing next to each other. For example, for the row $GBBGGGBGBGGGBGBGGBGG$ we have $S = 12$. What is the expected value of S for a randomly chosen arrangement of the 20 people? *(Source: American High School Mathematics Examination, now called the AMC 12)*

Solution for Problem 3.5: First, we note that we can treat all the boys as the same, and all the girls as the same. This reduces our problem to designating 7 spots as B (boys' spots) and 13 as G (girls' spots). We then must find the expected value of the number of times a B is next to a G.

We could start with a tedious straightforward approach – look at each possible arrangement and count the number of BG's and GB's for each:

Arrangement	# of BG's and GB's
$BBBBBBBGGGGGGGGGGGGG$	1
$BBBBBBGBGGGGGGGGGGGG$	3
$BBBBBBGGBGGGGGGGGGGG$	3
$BBBBBBGGGBGGGGGGGGGG$	3
$BBBBBBGGGGBGGGGGGGGG$	3

Um, this is going to take forever. Since there are $\binom{20}{7}$ = (a very big number) ways to choose the B places, our list will be very, very long. We need a better way.

Instead of looking at the problem 'arrangement-by-arrangement', we take an 'element-by-element' approach. We look at the first two spots and calculate in what portion of the arrangements they will be BG or GB. Clearly there are 2 ways to make the first two places a boy next to a girl. We must choose 6 of the remaining 18 places to have boys in them, so there are $\binom{18}{6}$ ways to allocate the rest of the places to boys or girls. Therefore, the number of arrangements in which the first two places have a boy and a girl is $2\binom{18}{6}$.

There's nothing special about the first 2 places! By the same reasoning, there are also $2\binom{18}{6}$ ways the second and third places can have one boy and one girl. And same for the third and fourth, and fourth and fifth, and so on! There are 19 such pairs (not 20), so the total number of times we will have a boy next to a girl is $19 \times 2\binom{18}{6}$. To get our expected value, we divide this by the total number of arrangements, which we found earlier is $\binom{20}{7}$:

$$\frac{19 \times 2\binom{18}{6}}{\binom{20}{7}} = \frac{19 \times 2 \times 18! \times 7! \times 13!}{6! \times 12! \times 20!} = \frac{91}{10}.$$

$\square$

Problem 3.6: Suppose you and I play a game. You start with 500 dollars and I start with 1000 dollars. We flip a fair coin repeatedly. Each time it comes up heads, I give you a dollar. Each time it comes up tails, you give me a dollar. We continue playing until one of us has all the money. What is the probability you will win this game?

Solution for Problem 3.6: This problem doesn't look like it has anything to do with expected value! But when we look at the problem through the lens of expected value, we find a slick solution.

The expected value of each coin toss is 0, because it is a fair coin. Since the expected value of each toss is 0, the expected value of any number of tosses is 0. Most importantly, no matter how long the game lasts, the expected value of the whole game is 0 for us both. Since your two possible outcomes are +$1000 and −$500, we can now write a simple expression for the expected value of the game for you:

$$E(\text{the game for you}) = (+\$1000)(p) + (-\$500)(1 - p).$$

Since your expected value is 0, we have $0 = (+\$1000)(p) + (-\$500)(1 - p)$, so $p = 1/3$. Therefore, the probability you will win the game is 1/3. □

Expected value can also be cleverly used for existence problems. For example, suppose that the expected value of the American Invitational Mathematics Exam (AIME) score of a USAMTS Gold Prize winner chosen at random is 9.4. From this, we can deduce that at least one of the Gold Prize winners scored a 10 or higher, and that at least one of them scored a 9 or lower. Therefore, expected value can sometimes be used to solve existence problems of the variety, "Show that there exists one item with at least ... " or "Show that there exists one item with at most ...."

We'll illustrate this strategy with an example.

Problem 3.7: There are 650 special points inside a circle of radius 16. You have a flat washer in the shape of an annulus (the region between two concentric circles), which has an inside radius of 2 and an outside radius of 3. Show that it is always possible to place the washer so that it covers up at least 10 of the special points. *(Source: Problem Solving Strategies by Arthur Engel. Solution by Ravi Boppana, USAMTS supporter)*

Solution for Problem 3.7: This is an existence problem; we wish to show that there exists some position at which we can place the washer so that it covers at least 10 points. This suggests an expected value approach. If we can show that the expected value of the number of points covered by a washer placed at random is greater than 9, then we can conclude that there must be some place we can put the washer to cover 10 or more points.

Calculating this expected value, however, doesn't appear so simple – for starters, there are infinitely many possible places to put the washer! In probability problems in which the possible range is continuous, we typically reach for the tools of geometry to measure the possible region and the desired region. So, we'll try that here. But how?

Evaluating each possible position of the washer will literally take forever, since there are an infinite number of places we can put it. However, there are only 650 points in the circle we have to consider. Therefore, we take an element-by-element approach as we did in the BG-arrangements problem. We calculate the portion of the washer placings in which each special point is covered.

Let G be one of our special points. G is covered by the washer if the center of the washer is at least 2 units away from the point, but no more than 3 units away. For example, the center of the washer shown is in our 'success' region, so the washer covers our point G. Any washer centered in the light grey region will cover point G.

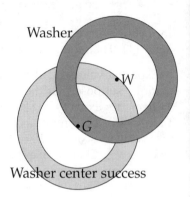

Washer

Washer center success

Therefore, the area of the region in which the center of the washer must be in order to cover G is $3^2\pi - 2^2\pi = 5\pi$.

We must be careful to include those cases in which the center of the washer is just outside our initial circle, but still covering special points near the circumference of the circle, as shown at right.

To take into account the possibility of a washer centered outside the circle covering special points inside the circle (an example of which is shown at right), we note that the 'possible' region in which the center of our washer can be placed and still have the washer cover special points is a circle with radius 19, not 16. Now, we can evaluate the probability that a washer placed at random covers a given special point. Out of the $19^2\pi = 361\pi$ area in which we can place the center of our washer to cover portions of our circle with the washer, there is a 5π area in which it can be placed to cover G. Therefore, the probability G is covered is $(5\pi)/(361\pi) = 5/361$.

There's nothing special about the G we examined! For each of the 650 special points, the probability that a randomly placed washer covers the point is 5/361. Therefore, each special point contributes 5/361 to the expected value of the number of special points covered by a randomly placed washer. So, the total expected value of the number of special points covered by a randomly placed washer is

$$650 \cdot \frac{5}{361} \approx 9.003.$$

Since the expected value of the number of special points covered by a randomly placed washer that overlaps some portion of our circle is greater than 9, there must be some placement of the washer which covers at least 10 special points. □

Art of Problem Solving Foundation
www.artofproblemsolving.org

Promoting problem solving education for middle and high school students.

The current programs administered through the Foundation are:
- **The USA Mathematical Talent Search**. See page 28 for more information about this free math competition.
- **The Local Programs Initiative**
- **Math Teams - San Diego**

Local Programs Initiative

This Foundation program provides an opportunity for local math clubs, math teams, and math circles to raise funds for speakers, activities, prizes, travel to competitions and entry fees, and more. By applying to be a part of the Local Programs Initiative, a local math club can accept tax-deductible donations from individuals, foundations, and corporations. An application can be found on the website: www.artofproblemsolving.org.

The Teacher's Circle   San Diego Math Circle

YOUR MATH CLUB HERE!

STANFORD MATH CIRCLE   Colorado ARML

Charlotte Mathematics Club

Soli Deo Gloria Home Educators

Math Teams - San Diego

This is a pilot program designed to help build math teams in the San Diego area. The Foundation will provide grants and training to middle school teachers in San Diego County to develop math teams at their schools and to participate in local and regional math events.

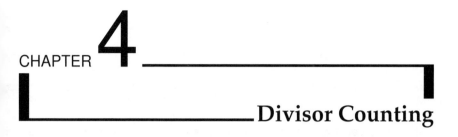

CHAPTER **4**

_____ **Divisor Counting**

Note: *This chapter is excerpted, with modifications, from the book* Introduction to Number Theory *by Mathew Crawford.*

4.1 Introduction

In this chapter we discuss a method (and develop a formula) for counting the number of positive divisors of an integer. We then use this method flexibly to count divisors that possess certain properties (multiples of 2, squares, cubes, etc.) and solve other problems.

While working the problems in this chapter, remember that your goal is to find ways to *organize* divisors of integers so that you can use their properties in creative ways to solve problems.

4.2 Counting Divisors

Problem 4.1: How many positive integers are divisors of 200?

Solution for Problem 4.1:

$$200 = 2^3 \cdot 5^2$$

A divisor of 200 can only have prime divisors of 2 and 5. This means that we can express any positive divisor of 200 in the form $2^a \cdot 5^b$ where a

must be an integer no greater than 3 and b must be an integer no greater than 2. The possible values of a are 0, 1, 2, and 3. The possible values of b are 0, 1, and 2.

$a \setminus b$	0	1	2
0	$2^0 \cdot 5^0 = 1$	$2^0 \cdot 5^1 = 5$	$2^0 \cdot 5^2 = 25$
1	$2^1 \cdot 5^0 = 2$	$2^1 \cdot 5^1 = 10$	$2^1 \cdot 5^2 = 50$
2	$2^2 \cdot 5^0 = 4$	$2^2 \cdot 5^1 = 20$	$2^2 \cdot 5^2 = 100$
3	$2^3 \cdot 5^0 = 8$	$2^3 \cdot 5^1 = 40$	$2^3 \cdot 5^2 = 200$

Each combination of values for a and b represents one divisor of 200 that we can construct. Since there are 4 values for a to choose from and 3 values for b to choose from, there are $4 \cdot 3 = 12$ total positive divisors of 200. In other words, we compute the number of positive divisors by taking the product of one more than each exponent in the prime factorization of 200.

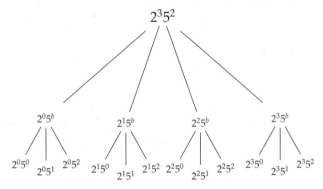

Now that we know the possible values of the exponents in the prime factorizations of divisors of 200, this tree diagram helps us visualize a method for counting them. We visually count 4 groups of 3 divisors and see that there are $3 \cdot 4 = 12$ divisors. □

> **Concept:** While counting divisors is not all that difficult for small natural numbers, it quickly becomes a chore for larger numbers. Understanding the relationships between the prime factorization of an integer and each of its divisors help us count the divisors more easily.

> **Important:** Since giving a quantity a name makes it easier to talk about, we denote the number of positive divisors of a natural number n as $t(n)$.

Problem 4.2: How many positive integers are divisors of 192?

Solution for Problem 4.2:

$$192 = 2^6 \cdot 3^1$$

Next, we note that any positive divisor of 192 must be in the form

$$2^a \cdot 3^b,$$

where a is 0, 1, 2, 3, 4, 5, or 6 and b is 0 or 1.

There are 7 possible values for a and 2 possible values for b, so $t(192) = 7 \cdot 2 = 14$. □

Problem 4.3: Let n be a natural number with prime factorization,

$$n = p_1{}^{e_1} \cdot p_2{}^{e_2} \cdots p_m{}^{e_m}.$$

Find a formula for the number of positive divisors of n.

Solution for Problem 4.3: Let's imagine how we can construct each positive divisor of n. The prime factorization of n includes anywhere from 0 to e_1 powers of p_1, so there are $e_1 + 1$ choices for the exponent of p_1 in the prime factorization of a divisor. Likewise, there are $e_2 + 1$ choices for the exponent of p_2, there are $e_3 + 1$ choices for the exponent of p_3, ..., and $e_m + 1$ choices for the exponent of p_m. All these choices correspond to the possible sets of exponents in the prime factorizations of the divisors of n. From these numbers of choices, we compute $t(n)$, the total number of positive divisors of n:

$$t(n) = (e_1 + 1)(e_2 + 1) \cdots (e_m + 1).$$

In other words, we find the total number of positive divisors of a natural number by taking the product of 1 more than each of the exponents in its prime factorization. □

> **Important:** Let n be a natural number with prime factorization,
>
> $$n = p_1{}^{e_1} \cdot p_2{}^{e_2} \cdots p_m{}^{e_m}.$$
>
> From this general prime factorization, we derived a handy formula for the number of positive divisors, $t(n)$, of a natural number n:
> $$t(n) = (e_1 + 1)(e_2 + 1) \cdots (e_m + 1).$$

> **Concept:** While it is convenient to know mathematical formulas (such as the one we derived for counting divisors), it is far more important to understand the method used. Over the course of this chapter we examine many divisor counting problems (where the formula we developed does not apply) that require creative solutions involving the *method* we developed.

Exercises for Section 4.2

4.2.1 Find the number of positive divisors of each integer using its prime factorization.

(a)	4	(e)	25	(i)	180
(b)	6	(f)	30	(j)	280
(c)	12	(g)	60	(k)	441
(d)	15	(h)	124	(l)	504

4.2.2 Find the number of positive divisors of 2002.

4.2.3 Karlanna places 600 marbles into m total boxes such that each box contains an equal number of marbles. There is more than one box, and each box contains more than one marble. For how many values of m can this be done?

4.3 Divisor Counting Problems

There are many kinds of divisor counting problems aside from the task of counting all of the positive divisors of an integer. These problems require not simply knowledge of the formula for counting positive divisors, but a *full understanding of the divisor counting process* we used to create the formula.

> **Problem 4.4:** How many of the positive divisors of 168 are even?

> **Bogus Solution:** The total number of positive divisors of 168 is $t(168) = 16$. Half of all integers are even, so half of the positive divisors of 168 are even, so there are 8 even positive divisors of 168.

Just because half of all integers are even doesn't mean that half of a particular group of integers is even. As the following solution shows us, more than half of the positive divisors of 168 are even.

Solution for Problem 4.4: It's usually helpful in a divisor problem to start with prime factorization:

$$168 = 2^3 \cdot 3^1 \cdot 7^1.$$

We know that each positive divisor of 168 has a prime factorization in the form $2^a \cdot 3^b \cdot 7^c$ where a is a whole number no greater than 3, where b is a whole number no greater than 1, and where c is a whole number no greater than 1.

We want to count only the even divisors, which are the ones with prime factorizations where a is at least 1. This leaves us with 3 choices for a (1, 2, or 3), and 2 choices for each of b and c (0 or 1). Multiplying the numbers of choices for each exponent in the prime factorization of an even divisor of 168, we find that the total number of even divisors of 168 is $3 \cdot 2 \cdot 2 = 12$.

> **Concept:** A second solution to this problem that involves a technique we call "complementary counting." **Complementary counting** is the process of counting what we do *not* want (instead of what we do want) and subtracting what we *don't* want from the whole:
>
> Total things–things we don't want = things we do want.

To use complementary counting to count the positive even divisors of 168, we first count the total number of positive divisors (16) and then subtract from that the total number of odd divisors. A divisor of 168 must have the form

$$2^a \cdot 3^b \cdot 7^c,$$

where a is 0, 1, 2, or 3, b is 0 or 1, and c is 0 or 1. In order for a divisor to be odd, a must be 0. This leaves 1 choice for a, 2 choices for b, and 2 choices for c, thus there are $1 \cdot 2 \cdot 2 = 4$ odd positive divisors of 168. Subtracting the number of odd divisors from the total gives us $16 - 4 = 12$ positive even divisors. $\square$

> **Problem 4.5:** Show that any positive perfect square has an odd number of positive divisors.

Solution for Problem 4.5: Let's start by examining the prime factorization of a perfect square. Starting with a general prime factorization of a positive integer n,

$$n = p_1^{e_1} \cdot p_2^{e_2} \cdots p_m^{e_m},$$

we square both sides to get a perfect square on the left, and a prime factorization with only even exponents on the right:

$$n^2 = p_1^{2e_1} \cdot p_2^{2e_2} \cdots p_m^{2e_m}.$$

Now we see that the number of positive divisors of the perfect square is

$$(2e_1 + 1)(2e_2 + 1) \cdots (2e_m + 1),$$

which is odd because it is the product of odd integers. $\square$

On a similar note, the prime factorization for a natural number that is not a perfect square must include some prime number raised to an odd exponent. If there were only even exponents then the number would be the square of some other number.

When there is at least one odd exponent, there is at least one prime for which there is an even number of choices for exponents with which to construct divisors. When we multiply together the number of choices for each exponent, the product must therefore also be even.

An example is $108 = 2^2 \cdot 3^3$, which has $(2 + 1)(3 + 1) = 12$ positive divisors. There are 3 powers of 3 in the prime factorization of 108 meaning that there are $3 + 1 = 4$ ways to choose the exponent of 3 when building a list of divisors, so the total number of divisors must be even:

$2^0 \cdot 3^0 = 1$	$2^0 \cdot 3^1 = 3$	$2^0 \cdot 3^2 = 9$	$2^0 \cdot 3^3 = 27$
$2^1 \cdot 3^0 = 2$	$2^1 \cdot 3^1 = 6$	$2^1 \cdot 3^2 = 18$	$2^1 \cdot 3^3 = 54$
$2^2 \cdot 3^0 = 4$	$2^2 \cdot 3^1 = 12$	$2^2 \cdot 3^2 = 36$	$2^2 \cdot 3^3 = 108$

Another way to see that perfect squares have an odd number of positive divisors and non-squares have an even number is to notice that for a non-square number n, every divisor d has a corresponding divisor a such that

$$n = da.$$

We know that d and a cannot be equal, otherwise n would be the square of either of them. Now, since we can always pair a d with an a, the total number of divisors can be counted in pairs! Thus the number of positive divisors is even. An example of this counting by pairs can be seen in the number 20, which is not a perfect square:

$$1 \cdot 20 = 20 \qquad 2 \cdot 10 = 20 \qquad 4 \cdot 5 = 20$$

The corresponding pairs d and a are 1 and 20, 2 and 10, and 4 and 5, thus the positive divisors of 20 can be counted in pairs.

These pairs can also be found in perfect squares. The only difference is that perfect squares have a single leftover divisor that cannot be paired with a **different** divisor. That divisor is the square root of the perfect square. Thus when we count up the divisors, we first count all the pairs and then add one more to get an odd number of divisors.

An example of this can be seen in the perfect square 36:

$$1 \cdot 36 = 36 \qquad 2 \cdot 18 = 36 \qquad 3 \cdot 12 = 36 \qquad 4 \cdot 9 = 36 \qquad 6^2 = 36$$

The corresponding pairs d and a are thus 1 and 36, 2 and 18, 3 and 12, and 4 and 9. The square root of 36 is 6, which is the unpaired divisor, bringing us to an odd total of 9 divisors.

> **Important:** A positive integer is a perfect square if and only if it has
> an odd number of positive divisors.

Recognizing this fact helps us determine useful information about integers.

> **Problem 4.6:** How many divisors of 5400 are **not** multiples of any
> perfect square greater than 1?

Solution for Problem 4.6: As in other divisor counting problems, we begin with the prime factorization:

$$5400 = 2^3 \cdot 3^3 \cdot 5^2.$$

We now consider what the factorization of a divisor, d, of 5400 looks like:

$$d = 2^a \cdot 3^b \cdot 5^c.$$

Our goal is to count only those positive divisors of 5400 that are not multiples of any perfect squares greater than 1, so we investigate the features of the prime factorization of such a divisor. When a, b, or c is greater than 1, d will be a multiple of the square of one of the primes in the factorization of 5400. For instance, if $a = 3$, then d is a multiple of $2^3 = 8$ and therefore a multiple of 4, a perfect square.

On the other hand, if each of a, b and c is no greater than 1 then none of 2^2, 3^2 or 5^2 are divisors of d. For instance, if $d = 2^1 \cdot 3^1 \cdot 5^1 = 30$, there is no prime whose square is a divisor of d. This means that no perfect square divides 30.

Now, there are 2 possible choices for each of a, b, and c which produce divisors that are not multiples of any squares greater than 1. This means there are $2 \cdot 2 \cdot 2 = 8$ ways to construct d such that d is not the multiple of a perfect square greater than 1. These constructed prime factorizations represent the 8 "square free" divisors of 5400. □

The solutions to the divisor counting problems in this chapter show how useful prime factorization can be. They also show the power of organizing information in a common mathematical form. Once we describe the possible prime factorizations of divisors, the counting part is relatively easy.

> **Concept:** Organizing information mathematically helps you more
> easily work toward a solution to a problem.

In the last problem we used prime factorization as the basis for our model. We then molded that model to count divisors that were not mutiples of any square greater than 1 by examining the properties of the exponents in the prime factorizations of such divisors.

Problem 4.7: If n has exactly 7 positive divisors, how many positive divisors does n^2 have?

Solution for Problem 4.7: We know that the total number of positive divisors of n is the product of one more than each exponent in the prime factorization of n. We know this product is equal to 7. Since 7 is prime, there can be only one number in this product, so the prime factorization of n must be $n = p^6$, for some prime number p.

We square the prime factorization of n to give us a prime factorization for n^2:

$$n^2 = (p^6)^2 = p^{12}.$$

Now we see that n^2 has $12 + 1 = 13$ positive divisors. $\square$

Problem 4.8: If n is the perfect cube of a natural number and n has 28 positive divisors, 2 of which are prime, how many positive divisors does the cube root of n have?

Solution for Problem 4.8: If possible, we would like to find a structure for the prime factorization of n. We can use the information we have about the number of primes and divisors of n to help us.

There are 28 combinations for the exponents in the prime factorization of a divisor of n. This equals the product of the numbers of choices for each exponent in the prime factorization of a divisor of n. Since n has only two prime divisors, we consider the ways in which we can break 28 down into the product of two smaller integers:

$$t(n) = 28 = 2 \cdot 14 = 4 \cdot 7.$$

Since each of these smaller integers represents the number of choices for possible exponents in the prime factorization of a divisor, we can determine

---

possible exponents in the prime factorization of n by subtracting 1 from each.

$$2, 14 \quad \rightarrow \quad 1, 13$$
$$4, 7 \quad \rightarrow \quad 3, 6$$

We now apply what we know about perfect cubes to the problem. We know that the exponents in the prime factorization of a perfect cube are multiples of 3. Since 1 and 13 are not multiples of 3, but 3 and 6 are, the exponents in the prime factorization of n must be 3 and 6. An example of a possible value of n is

$$2^6 \cdot 3^3 = 12^3 = 1728.$$

In general, we can say that the prime factorization of n looks like

$$n = p_1^6 \cdot p_2^3 = (p_1^2 \cdot p_2^1)^3.$$

Now we see that

$$\sqrt[3]{n} = p_1^2 \cdot p_2^1,$$

so the cube root of n has $(2 + 1)(1 + 1) = 6$ positive divisors. $\square$

> **Concept:** We first used prime factorizations of integers to help us determine divisor counts. Conversely, divisor counts themselves sometimes help us determine possible prime factorizations of integers. Understanding this relationship between the prime factorizations and divisor counts of integers gives us greater flexibility in solving problems.

Problem 4.9: How many of the positive divisors of 960 have 6 positive divisors?

Solution for Problem 4.9: The problem requires us to find divisors of 960 with a particular property, so we begin by finding the prime factorization of 960 in order to learn more about its divisors:

$$960 = 2^6 \cdot 3^1 \cdot 5^1.$$

A divisor d of 960 has the form

$$d = 2^a \cdot 3^b \cdot 5^c.$$

We must now count possible combinations of a, b and c such that d has exactly 6 positive divisors. In other words, we are counting combinations of a, b and c such that

$$(a + 1)(b + 1)(c + 1) = 6.$$

The only ways to get 6 as the product of 3 positive integers are

$$6 = 1 \cdot 1 \cdot 6 = 1 \cdot 2 \cdot 3.$$

These products imply that (a, b, c) must include either two 0's and one 5 or else one 0, one 1, and one 2. Since a is no greater than 6 and b and c are no greater than 1, this leaves us with only three possibilities: $(5, 0, 0)$, $(2, 0, 1)$, and $(2, 1, 0)$.

We can check these solutions to be sure that we found the divisors we wish to count:

$$\begin{aligned}
t(32) &= t(2^5) &= 6 \\
t(12) &= t(2^2 \cdot 3^1) &= 6 \\
t(20) &= t(2^2 \cdot 5^1) &= 6
\end{aligned}$$

So, the divisors of 960 which themselves have 6 positive divisors are 12, 20, and 32. $\square$

> **Concept:** Check your solutions to be sure that they solve the problem. This is even more important with unusual or difficult problems. So far, no math student was born who didn't make mistakes, but good problem solvers catch many of their own errors.

Problem 4.10: If m has 10 positive divisors, n has 6 positive divisors, and $\gcd(m, n) = 1$, how many positive divisors does mn have?

Solution for Problem 4.10: This problem is not as difficult as it might first appear. The real challenge is thinking creatively about m and n because we don't have any concrete numbers to work with.

We must think about what we know about m and n that could help us. We know that the two integers are relatively prime. This means that their prime factorizations have no common primes. In order to count the number of divisors in mn, we note that the prime factorization of mn is simply the prime factorizations of m and n multiplied together. We already know that the product of each exponent plus one in the factorization of m is 10 and that the product of each exponent plus one in the factorization of n is 6. This means that the product of one more than each exponent in the prime factorization of mn is $6 \cdot 10 = 60$.

An example of two such integers is $80 = 2^4 \cdot 5^1$ and $63 = 3^2 \cdot 7^1$. The prime factorization of the product involves simply relisting all of the primes and exponents in the two integers themselves:

$$80 \cdot 63 = 2^4 \cdot 3^2 \cdot 5^1 \cdot 7^1.$$

$\square$

By the same reasoning, if m and n are any two natural numbers such that $\gcd(m, n) = 1$,
$$t(mn) = t(m) \cdot t(n).$$

Exercises for Section 4.3

4.3.1 How many of the positive divisors of 252 are even?

4.3.2 How many of the positive divisors of 2160 are multiples of 3?

4.3.3 How many ordered pairs, (x, y), of positive integers satisfy the equation $xy = 144$?

4.3.4 How many positive divisors do 48 and 156 have in common?

4.3.5 What proportion of the positive divisors of 840 are prime?

4.3.6 Let n be an odd integer with exactly 11 positive divisors. Find the number of positive divisors of $8n^3$.

4.3.7 If $\gcd(m, n) = 1$, m has 8 positive divisors, and n has 12 positive divisors, how many positive divisors does mn have?

4.3.8

(a) How many positive divisors do 840 and 960 have in common?

(b) How many positive divisors do 840 and 1200 have in common?

(c) How many positive divisors do 960 and 1200 have in common?

(d) How many positive divisors do 840, 960, and 1200 have in common?

(e) How many natural numbers are positive divisors of exactly 2 of the 3 integers 840, 960, and 1200?

4.3.9 Let n be a natural number with exactly 2 positive prime divisors. If n^2 has 27 divisors, how many does n have?

4.3.10 How many of the positive divisors of 45000 themselves have exactly 12 positive divisors?

4.3.11 Let n be an integer with an odd number of positive divisors. Explain why $36n$ must also have an odd number of positive divisors.

4.4 Summary

> **Concept:** Prime factorization provides us with an organized approach to divisor counting problems.

While we can simply hunt down all the divisors of small natural numbers, the process quickly becomes tedious for larger integers. An organized approach provides us with flexibility and efficiency. Prime factorization is a great tool for organizing divisors and helped us derive the following formula:

> **Important:** When n is a positive integer with prime factorization
> $$n = p_1^{e_1} \cdot p_2^{e_2} \cdots p_m^{e_m},$$
> the formula for $t(n)$, the number of positive divisors of n, is
> $$t(n) = (e_1 + 1)(e_2 + 1) \cdots (e_m + 1).$$

> **Concept:** It is easier to talk about a quantity if we give it a name.

As mathematics gets more complicated, we continually refine the language we use to discuss mathematical ideas. We used $t(n)$ to denote the number of positive divisors of a positive integer n. This made it easier to discuss some of the problems in Chapter 4.

> **Concept:** While it is convenient to know the formulas for problems such as divisor counts, it is far more important to understand the *methods* involved.

Over the course of this chapter we tackled many problems that required us to examine the forms of the prime factorizations of natural numbers and then think about the ways we could construct different types of divisors. We could not simply plug numbers into a simple formula to solve all of these problems. Understanding the method helps us more easily solve a wide variety of divisor problems, including our proof of the following theorem:

> **Important:** A positive integer is a perfect square if and only if it has an odd number of positive divisors.

> **WARNING!!** Check all the assumptions that you make when solving a problem.

Some students might assume that half of the divisors of an integer are even because half of all integers are even. As we saw in Problem 4.4, this is not the case. Learning and exploring mathematics (and all other forms of problem solving) involves thinking a little deeper about facts that we take for granted. Checking our assumptions helps us refine our thinking processes.

> **Concept:** It is sometimes easier to count the things we don't want than the things we do want. **Complementary counting** is the process of counting what we do not want (instead of what we do want) and subtracting what we don't want from the whole:
>
> Total things−things we don't want = things we do want.

Complementary counting provided us with one solution to Problem 4.4. This problem solving technique is discussed more thoroughly in the book *Introduction to Counting & Probability*.

> **Concept:** Organizing information mathematically helps you more easily work toward a solution to a problem.

We found a formula for divisor counts by organizing divisors according to their prime factorizations.

> **Concept:** We first used prime factorizations of integers to help us determine divisor counts. Conversely, divisor counts themselves sometimes help us determine possible prime factorizations of integers. Understanding this relationship between the prime factorizations and divisor counts of integers gives us greater flexibility in solving problems.

> **Concept:** Check your solutions – particularly those to unusual or difficult problems.

Everybody makes mistakes. Catching your own mistakes means finding the solutions to more problems. Checking your work carefully means reviewing your own understanding of the methods you use and gives you more experience learning how to avoid making errors.

> **Concept:** Identifying the differences between an unusual or difficult problem and more ordinary problems often helps us determine how we can use the methods for solving the ordinary problems to solve the more unusual and difficult problems.

It pays to think about what is unusual, difficult, or interesting about a problem. Identifying the particular features of a problem allows you to more easily divide a problem into simpler parts – just like the problem parts at the beginning of each section in this book.

Review Problems for Chapter 4

4.11 Find the number of positive divisors of each of the following.

(a) 18   (c) 216   (e) 2520

(b) 52   (d) 420   (f) 3750

4.12 Find the number of distinct positive divisors of $(30)^4$ excluding 1 and $(30)^4$. *(Source: AHSME)*

4.13 Joseph and Timothy play a game in which Joseph picks an integer between 1 and 1000 inclusive and Timothy divides 1000 by that integer and states whether or not the quotient is an integer. How many integers could Joseph pick such that Timothy's quotient is an integer?

4.14 How many of the positive divisors of 3240 are multiples of 3?

4.15 How many of the divisors of 3240 are perfect squares?

Challenge Problems for Chapter 4

4.16 How many positive integers are divisors of 999,999?

4.17 Let m and n be two relatively prime natural numbers. If m has 12 positive divisors and n has 10 positive divisors, what is the product of the positive divisors of mn?

4.18 The total number of positive divisors of an integer is prime. How many prime numbers are divisors of the integer?

4.19 What is the sum of the three numbers less than 1000 that have exactly five positive integer divisors? *(Source: MATHCOUNTS)*

4.20

(a) Find the prime factorization of 3200.

(b) How many positive divisors does 3200 have?

(c) How many of the positive divisors of 3200 are even?

(d) How many of the positive divisors of 3200 are perfect squares?

(e)   How many of the positive divisors of 3200 are not multiples of any perfect square greater than 1?

4.21   If n has 2 prime divisors and 9 total divisors, how many divisors does n^2 have?

4.22   Find the smallest positive integer that can be written as a product of two factors (each greater than 1) in exactly three different ways. For example, 70 is one such integer since $70 = 2 \cdot 35 = 5 \cdot 14 = 7 \cdot 10$. Note that the order of the factors does not matter. *(Source: Mandelbrot)*

4.23   How many of the positive divisors of 14400 themselves have exactly 8 positive divisors?

4.24   How many positive divisors do 8400 and 7560 have in common?

4.25   A certain integer has 20 positive divisors.

(a)   What is the smallest number of primes that could divide the integer?

(b)   What is the largest number of primes that could divide the integer?

(c)   What is the smallest natural number that has exactly 20 positive divisors?

(d)★   Are there any natural numbers smaller than your answer from (c) that have more than 20 positive divisors?

4.26   For some natural number n, the integer $6n$ has 9 positive divisors.

(a)   How many prime numbers are factors of $6n$?

(b)   What is the exponent of each of those primes in the prime factorization of $6n$?

(c)   What is n?

4.27   What is the sum of all positive integers less than 100 that have exactly twelve divisors? *(Source: Mandelbrot)*

4.28★   Find the sum of the perfect square divisors of the smallest integer with exactly 6 perfect square divisors.

4.29★   How many positive integers have exactly three proper divisors, each of which is less than 50? *(Source: AIME)*

Art of Problem Solving
Online School

Subject Classes: Designed for strong math students in grades 7-12.

- **Introduction to Counting & Probability*** (12 weeks)
- **Introduction to Geometry*** (18 weeks)
- **Introduction to Number Theory*** (12 weeks)
- **Intermediate Algebra** (12 weeks)
- **Intermediate Counting & Probability** (12 weeks)
- **Intermediate Number Theory Seminar** (8 weeks)
- **Intermediate Trigonometry/Complex Numbers** (12 weeks)

*These classes have an accompanying textbook.

Problem Series: These classes provide preparation for some of the most prominent middle and high school programs in the US.

- **MATHCOUNTS Problem Series** (12 weeks)
- **AMC 10 Problem Series** (12 weeks)
- **AMC 12 Problem Series** (12 weeks)
- **AIME Problem Series A and B** (12 weeks)
- **Special AIME Problem Seminar** (2 days/6 hours)

All classes take place in a live moderated chat room. A private message board is available for each class for students to ask questions and communicate with each other. Subject Classes may include problem sets. All classes include class transcripts that students may access during the course of each class. For a current schedule of classes, please visit www.artofproblemsolving.com.

Testimonials

"AoPS classes have definitely made me a better problem solver and have helped me expand my mathematics horizon as well as increase my score on tests such as the AMC/AIME."

"I had so much fun in this class. Problems that I never would have imagined doing on my own became trivial. Thank you so much for giving me this unique opportunity to learn."

"I learned more than I had ever dreamt of. I look forward to participating in other classes offered here. Thank you for all the time and energy you have put into this course."

CHAPTER 5

Similar Triangles

Note: This chapter is excerpted from the book Introduction to Geometry *by Richard Rusczyk.*

5.1 What is Similarity?

We call two figures **similar** if one is simply a blown-up, and possibly rotated and/or flipped, version of the other. Our first problem gives us an example of similar figures.

Problem 5.1:

(a) Use a ruler to approximate the following ratios in Figure 5.1:

$$\frac{AB}{EH} \qquad \frac{BC}{HG} \qquad \frac{CD}{GF} \qquad \frac{DA}{FE} \qquad \frac{BD}{HF} \qquad \frac{AC}{EG}$$

(b) Measure angles $\angle A$ through $\angle H$.

(c) Do you find anything interesting in your answers to the first two parts?

Figure 5.1: Two Similar Figures

Solution for Problem 5.1: Measuring each of the segments in the given ratios, we find that in each case, the ratio is 1/2. When we measure the angles, we find that the angles of *ABCD* are equal to those in *EHGF* (note the orders of the vertices!):

$$
\begin{aligned}
\angle A &= \angle E &= 90° \\
\angle B &= \angle H &= 130° \\
\angle C &= \angle G &= 80° \\
\angle D &= \angle F &= 60°
\end{aligned}
$$

□

We write the similarity in Figure 5.1 as *ABCD* ~ *EHGF* since ∠*A* corresponds to ∠*E*, ∠*B* corresponds to ∠*H*, etc. As with congruence, we have to be careful about the order of the vertices. For example, we would not write *ABCD* ~ *EFGH* to describe Figure 5.1.

The ratio between corresponding lengths in similar figures is constant, and is equal to the ratio by which one figure is 'blown up' to get the other. In Figure 5.1, we have

$$
\frac{AB}{EH} = \frac{BC}{HG} = \frac{CD}{GF} = \frac{DA}{FE}.
$$

All corresponding lengths of *ABCD* and *EHGF* follow this ratio. For example, we could include *BD/HF* and *AC/EG* in that chain of equalities above.

As we saw in Problem 5.1, corresponding angles in similar figures are equal.

Similar figures do not need to have the same orientation. The diagram to the right shows two similar triangles with different orientations.

Speaking of triangles, we'll be spending the rest of this chapter discussing how to tell when two triangles are similar, and how to use similar triangles once we find them. Below are a couple Exercises that provide practice using triangle similarities to write equations involving side lengths.

Exercises for Section 5.1

5.1.1 Given that $\triangle ABC \sim \triangle YXZ$, which of the following statements must be true:

(a) $AB/YX = AC/YZ$.

(b) $AB/BC = YX/XZ$.

(c) $AB/XZ = BC/YX$.

(d) $(AC)(YX) = (YZ)(BA)$.

(e) $BC/BA = XY/ZY$.

5.1.2 $\triangle ABC \sim \triangle ADB$, $AC = 4$, and $AD = 9$. What is AB? *(Source: MATHCOUNTS)*

5.2 AA Similarity

In our introduction, we stated that similar figures have all corresponding angles equal, and that corresponding sides are in a constant ratio. It sounds like a lot of work to prove all of that; however, just as for triangle congruence, we have some shortcuts to prove that triangles are similar. We'll start with the most commonly used method.

Important: **Angle-Angle Similarity (AA Similarity)** tells us that if two angles of one triangle equal two angles of another, then the triangles are similar.

$\angle A = \angle D$ and $\angle B = \angle E$ together imply $\triangle ABC \sim \triangle DEF$, so

$$\frac{AB}{DE} = \frac{AC}{DF} = \frac{BC}{EF}.$$

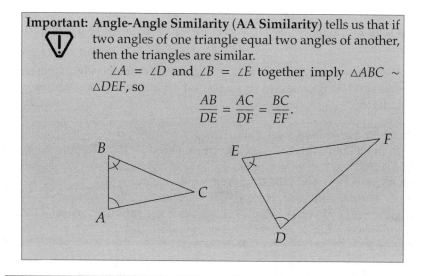

We'll explore why AA Similarity works in Section 5.5, but first we'll get some experience using it in some problems.

Problem 5.2: Below are two triangles that have the same measures for two angles.

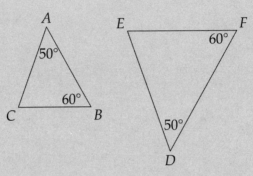

Find the third angle in each, and find the ratios AB/DF, AC/DE, BC/EF by measuring the sides with a ruler.

Solution for Problem 5.2: The last angle in each triangle is $180° - 50° - 60° = 70°$, so the angles of $\triangle ABC$ match those of $\triangle DFE$. In the same way, if we ever have two angles of one triangle equal to two angles of another, we know that the third angles in the two triangles are equal.

Measuring, we find that the ratios are each 1.5. It appears to be the case that if all the angles of two triangles are equal, then the two triangles are similar. □

We might wonder if two figures with equal corresponding angles are always similar. So, we add an angle and see if it works for figures with four angles.

Problem 5.3: Does your rule work for figures with more than 3 angles? Can you create a figure $EFGH$ that has the same angles as $ABCD$ at right such that $EFGH$ and $ABCD$ are not similar? (In other words, can you create $EFGH$ so that the angles of $EFGH$ equal those of $ABCD$, but the ratio of corresponding sides between $EFGH$ and $ABCD$ is not the same?)

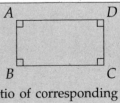

Solution for Problem 5.3: We can quickly find such an $EFGH$. The diagram to the right shows a square $EFGH$ next to our initial rectangle. Clearly these figures have the same angles, but when we check the ratios, we find that

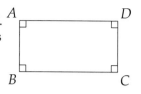 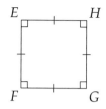

$$\frac{AB}{EF} < 1 < \frac{BC}{FG}.$$

$ABCD$ and $EFGH$ are not similar, so equal angles are not enough to prove similarity here. □

Let's return to triangles and tackle some problems using AA Similarity.

Problem 5.4: In the figure at right, $\overline{MN} \parallel \overline{OP}$, $OP = 12$, $MO = 10$, and $LM = 5$. Find MN.

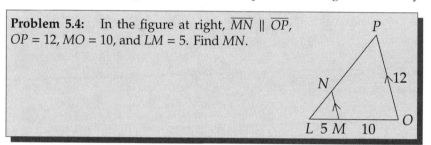

Solution for Problem 5.4: See if you can find the flaw in this solution:

Bogus Solution: Since $\overline{MN} \parallel \overline{OP}$, we have $\angle LMN = \angle LOP$ and $\angle LNM = \angle LPO$. Therefore, $\triangle LMN \sim \triangle LOP$, so $LM/MO = MN/OP$. Substituting our given side lengths gives $5/10 = MN/12$, so $MN = 6$.

Everything in this solution is correct except for $LM/MO = MN/OP$. $\overline{MO}$ is not a side of one of our similar triangles! The correct equation is $LM/LO = MN/OP$. Since $LO = LM + MO = 15$, we now have $5/15 = MN/12$, so $MN = 4$. □

Problem 5.5: The lengths in the diagram are as marked, and $\overline{WX} \parallel \overline{YZ}$. Find PY and WX.

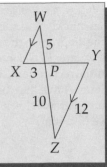

Solution for Problem 5.5: Where does this solution go wrong:

Bogus Solution: Since $\overline{WX} \parallel \overline{ZY}$, we have $\angle W = \angle Z$ and $\angle X = \angle Y$. Therefore, $\triangle WPX \sim \triangle YPZ$, and we have

$$\frac{PX}{PZ} = \frac{WX}{YZ} = \frac{WP}{PY}$$

Substitution gives

$$\frac{3}{10} = \frac{WX}{12} = \frac{5}{PY}.$$

We can now easily find $YP = 50/3$ and $WX = 18/5$.

This solution doesn't get the vertex order in the similar triangles right, so it sets up the ratios wrong! $\overline{PX}$ and $\overline{PZ}$ are not corresponding sides. $\overline{PX}$ in $\triangle WPX$ corresponds to $\overline{PY}$ in $\triangle ZPY$ because $\angle W = \angle Z$.

Here's what the solution should look like. Pay close attention to the vertex order in the similarity relationship.

Since $\overline{WX} \parallel \overline{ZY}$, we have $\angle W = \angle Z$ and $\angle X = \angle Y$. Therefore, $\triangle WPX \sim \triangle ZPY$. Hence, we have

$$\frac{PX}{PY} = \frac{WX}{YZ} = \frac{WP}{PZ}.$$

Substitution gives

$$\frac{3}{PY} = \frac{WX}{12} = \frac{5}{10}.$$

We can now easily find $PY = 6$ and $WX = 6$. □

Perhaps you see a common thread in the last two problems. While you won't always find parallel lines in similar triangle problems, you'll almost always find similar triangles when you have parallel lines.

> **Important:** Parallel lines mean equal angles. Equal angles mean similar triangles. The figures below show two very common set-ups in which parallel lines lead to similar triangles. Specifically, $\triangle PQR \sim \triangle PST$ and $\triangle JKL \sim \triangle MNL$.

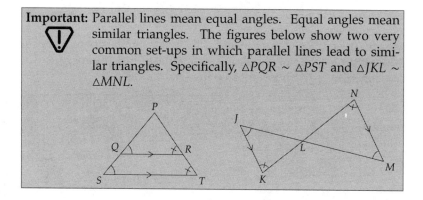

> **WARNING!!** Read the Bogus Solutions to Problems 5.4 and 5.5 again. These are very common errors; understand them so you can avoid them.

Problem 5.6: Find BC and DC given $AD = 3$, $BD = 4$, and $AB = 5$.

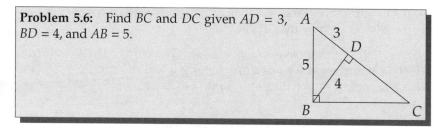

Solution for Problem 5.6: Since $\angle BAD = \angle CAB$ and $\angle BDA = \angle CBA$, we have $\triangle BAD \sim \triangle CAB$ by AA Similarity. Therefore, we have $BC/BD = AB/AD = 5/3$, so $BC = (5/3)(BD) = 20/3$.

We can use this same similarity to find AC, and then subtract AD to get CD. We could also note that $\angle BCD = \angle BCA$ and $\angle BDC = \angle CBA$, so $\triangle BCD \sim \triangle ACB$ by AA Similarity. Therefore, $CD/BD = BC/AB = (20/3)/5 = 4/3$, so $CD = (4/3)(BD) = 16/3$. $\square$

Similar triangles – they're not just for parallel lines.

> **Important:** Similar triangles frequently pop up in problems with right angles. The diagram in Problem 5.6 shows a common way this occurs. Make sure you see that
>
> $$\triangle ABD \sim \triangle BCD \sim \triangle ACB.$$

As you'll see throughout the rest of the book, similar triangles occur in all sorts of problems, not just those with parallel lines and perpendicular lines. They're also an important step in many proofs.

Problem 5.7: Given that $\overline{DE} \parallel \overline{BC}$ and $\overline{AY} \parallel \overline{XC}$, prove that

$$\frac{EY}{EX} = \frac{AD}{DB}.$$

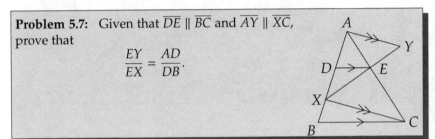

Solution for Problem 5.7: Parallel lines mean similar triangles. The ratios of side lengths in the problem also suggest we look for similar triangles.

Since $\overline{AY} \parallel \overline{XC}$, we have $\triangle AYE \sim \triangle CXE$. Now we look at what this means for our ratios. From $\triangle AYE \sim \triangle CXE$, we have $EY/EX = AE/EC$. All we have left is to show that $AE/EC = AD/DB$.

Since $\overline{DE} \parallel \overline{BC}$, we have $\triangle ADE \sim \triangle ABC$. Therefore, $AD/AB = AE/AC$, which is almost what we want! We break AB and AC into $AD + DB$ and $AE + EC$, hoping we can do a little algebra to finish:

$$\frac{AD}{AD + DB} = \frac{AE}{AE + EC}$$

If only we could get rid of the AD and AE in the denominators – then we would have $AD/DB = AE/EC$. Fortunately, we can do it. We can flip both fractions:

$$\frac{AD + DB}{AD} = \frac{AE + EC}{AE}.$$

Therefore, $\frac{AD}{AD} + \frac{DB}{AD} = \frac{AE}{AE} + \frac{EC}{AE}$, so $1 + \frac{DB}{AD} = 1 + \frac{EC}{AE}$, which gives us

$$\frac{DB}{AD} = \frac{EC}{AE}.$$

Flipping these fractions back over gives us $AD/DB = AE/EC$. Therefore, we have $EY/EX = AE/EC = AD/DB$, as desired. □

Our solution to the previous problem reveals another handy relationship involving similar triangles:

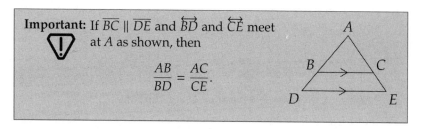

Important: If $\overline{BC} \parallel \overline{DE}$ and $\overleftrightarrow{BD}$ and $\overleftrightarrow{CE}$ meet at A as shown, then

$$\frac{AB}{BD} = \frac{AC}{CE}.$$

Exercises for Section 5.2

5.2.1

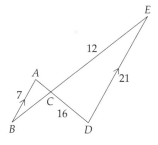

(a) Find AC and BC.

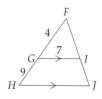

(b) Find HJ.

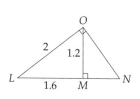

(c) Find ON and MN.

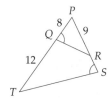

(d) Find RS.

5.2.2 If two isosceles triangles have vertex angles that have the same measure, are the two triangles similar? Why or why not?

5.2.3 In the diagram, $WXYZ$ is a square. M is the midpoint of $\overline{YZ}$, and $\overline{AB} \perp \overline{MX}$.

(a) Show that $\overline{WZ} \parallel \overline{XY}$.

(b) Prove that $AZ = YB$.

(c) Prove that $XB = XA$.

(d) Prove that $\triangle AZM \sim \triangle MYX$, and use this fact to prove $AZ = XY/4$.

5.2.4 In triangle ABC, $AB = AC$, $BC = 1$, and $\angle BAC = 36°$. Let D be the point on side $\overline{AC}$ such that $\angle ABD = \angle CBD$.

(a) Prove that triangles ABC and BCD are similar.

(b)★ Find AB.

5.2.5★ Find x in terms of y given the diagram below.

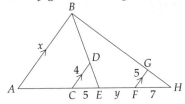

5.3 SAS Similarity

Problem 5.8:

(a) Measure $\overline{BC}$, $\overline{EF}$, and angles $\angle B$, $\angle C$, $\angle E$, and $\angle F$.

(b) Can you make a guess about how to use Side-Angle-Side for triangle similarity?

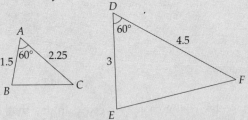

Solution for Problem 5.8: We aren't surprised to find that *BC* appears to be half *EF*: *BC* is about 2 cm and *EF* is around 4 cm. We also aren't shocked to find that ∠*B* appears to equal ∠*E* and ∠*C* appears to equal ∠*F*.

This example suggests that if two sides in one triangle are in the same ratio as two sides in another triangle (as *AB/AC = DE/DF*), and the angles between these sides are equal (as ∠*A* = ∠*D*), then the triangles are similar. □

No doubt, you know where this is headed. Time to develop a proof for our guess. As usual, we try to use what we already know, AA Similarity, to prove our guess for 'SAS Similarity'.

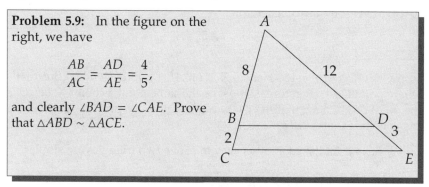

Problem 5.9: In the figure on the right, we have

$$\frac{AB}{AC} = \frac{AD}{AE} = \frac{4}{5},$$

and clearly ∠*BAD* = ∠*CAE*. Prove that △*ABD* ∼ △*ACE*.

Solution for Problem 5.9: What did we do wrong here:

> **Bogus Solution:** Since $\overline{BD} \parallel \overline{CE}$, we have ∠*ABD* = ∠*ACE* and ∠*ADB* = ∠*AEC*, so △*ABD* ∼ △*ACE* by AA Similarity.

There's not a single false statement in that solution. However, the assertion that $\overline{BD} \parallel \overline{CE}$ needs to be proved, and our Bogus Solution merely states it without justification.

In the solution below, we take the clever tactic of considering the point *X* on $\overline{AE}$ such that $\overline{BX} \parallel \overline{CE}$. Then we prove that *X* is in fact *D*.

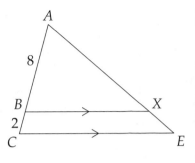

We'd like to prove that $\overline{BD} \parallel \overline{CE}$, but there's no obvious way to even start. We seem stuck, so we try to go a different direction. We create a point X on $\overline{AE}$ as shown at right, such that $\overline{BX} \parallel \overline{CE}$. Our goal now is to show that X must be D. Notice that we are not assuming that $\overline{BD} \parallel \overline{CE}$. We are taking some other point, X, such that $\overline{BX} \parallel \overline{CE}$, then trying to prove that X *must be* D.

Since $\overline{BX} \parallel \overline{CE}$, we have $\angle ABX = \angle ACE$ and $\angle AXB = \angle AEC$, so $\triangle ABX \sim \triangle ACE$ by AA Similarity. Therefore,

$$\frac{AX}{AE} = \frac{AB}{AC} = \frac{4}{5},$$

so $AX = (4/5)(AE) = 12$. Hence, X is on $\overline{AE}$ 12 units from A. But that's where point D is! Therefore, D must be the same point as point X; i.e., D is the point on $\overline{AE}$ such that $\overline{BD} \parallel \overline{CE}$. Now that we've proved $\overline{BD} \parallel \overline{CE}$, we can conclude that $\triangle ABD \sim \triangle ACE$. □

We have established another way to prove two triangles are similar.

> **Important: Side-Angle-Side Similarity (SAS Similarity)** tells us that if two sides in one triangle are in the same ratio as two sides in another triangle (as $AB/AC = DE/DF$ below), and the angles between these sides are equal (as $\angle A = \angle D$ below), then the triangles are similar.
>
>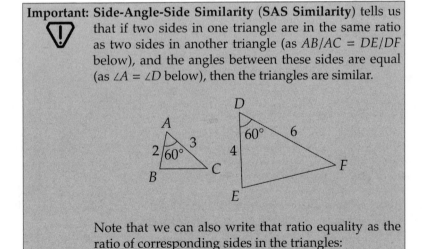
>
> Note that we can also write that ratio equality as the ratio of corresponding sides in the triangles:
> $AB/DE = AC/DF$.

You may be wondering how our solution to Problem 5.9 can be used to prove SAS Similarity in general, since Problem 5.9 only deals with the case of two triangles that share an angle, as $\triangle ABD$ and $\triangle ACE$ share $\angle A$. We can use this approach generally because if an angle in one triangle equals an angle in another, we can always slide (and/or flip) one triangle until it's on top of the other, as shown in Figure 5.2.

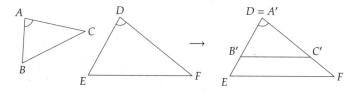

Figure 5.2: Sliding Triangles to Prove Similarity

SAS Similarity is most often used in diagrams like the one shown in Problem 5.9. However, it does come up in less obvious situations.

Problem 5.10: Given $AC = 4$, $CD = 5$, and $AB = 6$ as in the diagram, find BC if the perimeter of $\triangle BCD$ is 20. *(Source: Mandelbrot)*

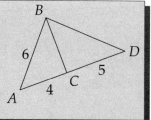

Solution for Problem 5.10: Since

$$\frac{AC}{AB} = \frac{4}{6} = \frac{2}{3} \quad \text{and} \quad \frac{AB}{AD} = \frac{6}{9} = \frac{2}{3},$$

we have $\triangle ACB \sim \triangle ABD$ by SAS (since the angle between the sides in each ratio above is $\angle A$). Since the sides of $\triangle ABD$ are 3/2 the corresponding sides of $\triangle ACB$, we have $BD = 3BC/2$. Now we can use that perimeter information. Since $BC + CD + DB = 20$, we have

$$BC + 5 + \frac{3BC}{2} = 20.$$

Therefore, $BC = 6$. $\square$

Exercises for Section 5.3

5.3.1 Find DE in the figure at left below.

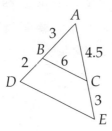
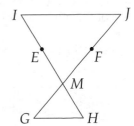

5.3.2 In the figure at right above, M is the midpoint of $\overline{EH}$ and of $\overline{FG}$. E and F are midpoints of $\overline{IM}$ and $\overline{MJ}$, respectively. Prove that $\overline{IJ} \parallel \overline{GH}$.

5.3.3 Show that if $WZ^2 = (WX)(WY)$ in the diagram at left below, then $\angle WZX = \angle WYZ$.

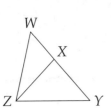
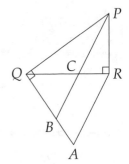

5.3.4 In the diagram at right above, $\angle PRQ = \angle PQA = 90°$, $QR = QA$, and $\angle QPC = \angle RPC$.

(a) Prove $\angle QCB = \angle QBC$.

(b)★ Prove $\overline{RA} \parallel \overline{PB}$.

5.4 SSS Similarity

We use SSS Similarity less often than AA and SAS.

Important: **Side-Side-Side Similarity (SSS Similarity)** tells us that
if each side of one triangle is the same constant multiple
of the corresponding side of another triangle, then the
triangles are similar. (And therefore, their corresponding
ing angles are equal.)

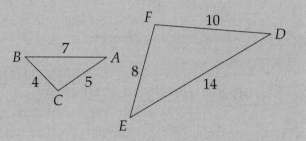

For example, in the diagram, we have

$$\frac{AB}{DE} = \frac{AC}{DF} = \frac{BC}{EF},$$

so

$$\triangle ABC \sim \triangle DEF.$$

Therefore, $\angle A = \angle D$, $\angle B = \angle E$, and $\angle C = \angle F$.

As we noted, few problems require SSS Similarity. We may, however,
consider it in problems in which all we are given is lengths, but we have
to prove something about angles.

Problem 5.11: Given the side lengths shown in
the diagram, prove that $\overline{AE} \parallel \overline{BC}$ and $\overline{AB} \parallel \overline{DE}$.

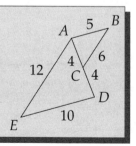

Solution for Problem 5.11: We need to use angles to show the segments are
parallel, but all we have are sides. We look for similarity, and see that

$$\frac{AB}{DE} = \frac{AC}{AD} = \frac{BC}{AE} = \frac{1}{2},$$

so $\triangle ABC \sim \triangle DEA$ by SSS Similarity. Therefore, $\angle BAC = \angle EDA$, so $\overline{AB} \parallel \overline{DE}$. Also, $\angle DAE = \angle ACB$, so $\overline{AE} \parallel \overline{BC}$. □

Exercises for Section 5.4

5.4.1 Two isosceles triangles have the same ratio of leg length to base length. Prove that the vertex angles of the two triangles are equal.

5.5 Using Similarity in Problems

In this section we explore some challenging problems that are solved with similar triangles, and we discover why AA Similarity works.

We start off with some warm-ups involving parallel and perpendicular lines.

Problem 5.12: In the diagram, $\overline{DE} \parallel \overline{BC}$, and the segments have the lengths shown in the diagram. Find x, y, and z.

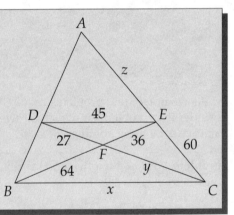

Solution for Problem 5.12: Since $\overline{ED} \parallel \overline{BC}$, we have $\triangle FBC \sim \triangle FED$ by AA Similarity. Therefore, we have

$$\frac{FC}{FD} = \frac{BC}{DE} = \frac{FB}{FE} = \frac{64}{36} = \frac{16}{9}.$$

Solving for x and y, we find $x = BC = (16/9)(DE) = 80$ and $y = FC = (16/9)(DF) = 48$.

Since $\triangle ADE \sim \triangle ABC$ by AA Similarity, we have $AE/AC = DE/BC =$

$45/80 = 9/16$. Since $AE = z$ and $AC = AE + EC = z + 60$, we have $z/(z + 60) = 9/16$. Cross-multiplying gives $16z = 9z + 540$, so $z = 540/7$. □

Problem 5.13: As shown in the diagram, $\angle A = 90°$, and $ADEF$ is a square. Given that $AB = 6$ and $AC = 10$, find AD.

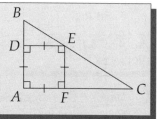

Solution for Problem 5.13: Since $\angle A = \angle EFC = 90°$, we have $\overline{EF} \parallel \overline{AB}$; similarly, $\overline{DE} \parallel \overline{AC}$. Therefore, this problem has both right triangles and parallel lines. Our parallel lines quickly tell us that by AA, we have

$$\triangle BDE \sim \triangle BAC \sim \triangle EFC.$$

If we let each side of $ADEF$ be x, we have $BD = 6 - x$ and $FC = 10 - x$. Our similar triangles can then be used to solve for x. From $\triangle BDE \sim \triangle EFC$, we have $BD/DE = EF/FC$. Substitution gives

$$\frac{6 - x}{x} = \frac{x}{10 - x}.$$

Cross-multiplying and solving the resulting equation for x gives $x = 15/4$. Therefore, $AD = x = 15/4$. □

Problem 5.14: In the diagram, $\overline{PX}$ is the altitude from right angle $\angle QPR$ of right triangle PQR as shown. Show that $PX^2 = (QX)(RX)$, $PR^2 = (RX)(RQ)$, and $PQ^2 = (QX)(QR)$.

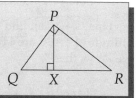

Solution for Problem 5.14: Right triangles mean similar triangles. $\angle PXR = \angle QPR$ and $\angle PRX = \angle PRQ$, so $\triangle PXR \sim \triangle QPR$. Therefore, we have $PR/RX = RQ/PR$, so $PR^2 = (RX)(RQ)$. Similarly, we can show $\triangle PQX \sim \triangle RQP$, so $PQ/QX = QR/PQ$, and we have $PQ^2 = (QX)(QR)$.

Combining the two triangle similarities (or by noting that $\angle XPQ = 90° - \angle XPR = \angle XRP$ and $\angle PXQ = \angle PXR$), we find $\triangle PXQ \sim \triangle RXP$. Therefore, $PX/QX = RX/PX$, so $PX^2 = (RX)(QX)$. □

The square root of the product of two numbers is called the **geometric mean** of the two numbers. The previous problem suggests where the name 'geometric mean' comes from. For example, what is the geometric mean of QX and RX?

Problem 5.15: Given that $\triangle ABC \sim \triangle XYZ$, $AB/XY = 4$, and $[ABC] = 64$, find $[XYZ]$.

Solution for Problem 5.15: Since $\triangle ABC \sim \triangle XYZ$ and $AB/XY = 4$, the ratio of corresponding lengths in the triangles is $4/1$. Therefore, the altitude of $\triangle ABC$ to $\overline{AB}$ is 4 times the corresponding altitude to $\overline{XY}$ in $\triangle XYZ$.

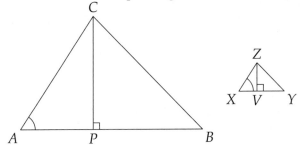

For a quick proof, consider the diagram above, in which we've drawn the aforementioned altitudes to $\overline{AB}$ and $\overline{XY}$. Since $\triangle ABC \sim \triangle XYZ$, we have $\angle A = \angle X$. Combining this angle equality with $\angle CPA = \angle ZVX$ gives $\triangle APC \sim \triangle XVZ$ by AA, so $CP/ZV = AC/XZ$. Since $\overline{AC}$ and $\overline{XZ}$ are corresponding sides of our original triangles, their ratio is $4/1$, so $CP/ZV = 4/1$.

Finally, we can find the ratio $[ABC]/[XYZ]$. Since both the base and the altitude of $\triangle ABC$ are 4 times the corresponding base and altitude of $\triangle XYZ$, we know that

$$[ABC]/[XYZ] = \frac{(AB)(CP)/2}{(XY)(ZV)/2} = \left(\frac{AB}{XY}\right)\left(\frac{CP}{ZV}\right) = \left(\frac{4}{1}\right)^2 = 16.$$

So, we have $[XYZ] = [ABC]/16 = 4$. $\square$

The same procedure we used to solve this problem can be used to find an important relationship between the areas of two similar triangles.

> **Important:** If two triangles are similar such that the sides of the
> larger triangle are k times the sides of the smaller, then
> the area of the larger triangle is k^2 times that of the
> smaller.
>
> This relationship holds for any pair of similar fig-
> ures, not just for triangles.

Problem 5.16: In the diagram, $\angle ACQ = \angle QCB$, $\overline{AQ} \perp \overline{CQ}$, and P is the midpoint of $\overline{AB}$. Prove that $\overline{PQ} \parallel \overline{BC}$.

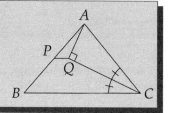

Solution for Problem 5.16: If we could show that $\angle QPA = \angle B$, then we could use that to prove $\overline{PQ} \parallel \overline{BC}$. Unfortunately, there are no obvious similar triangles or congruent triangles we can use to show that $\angle QPA = \angle B$.

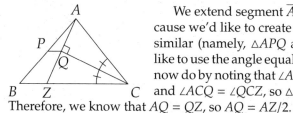

We extend segment $\overline{AQ}$ to point Z on $\overline{BC}$ because we'd like to create triangles that might be similar (namely, $\triangle APQ$ and $\triangle ABZ$). We'd also like to use the angle equalities at C, which we can now do by noting that $\angle AQC = \angle CQZ$, $CQ = CQ$, and $\angle ACQ = \angle QCZ$, so $\triangle CQZ \cong \triangle CQA$ by ASA. Therefore, we know that $AQ = QZ$, so $AQ = AZ/2$.

We might seem stuck here, but then we remember the last bit of information we haven't used. Since P is the midpoint of $\overline{AB}$, we have $AP = AB/2$, so $\triangle PAQ \sim \triangle BAZ$ by SAS Similarity. Thus, $\angle APQ = \angle B$, so $\overline{PQ} \parallel \overline{BC}$. □

> **Concept:** When you're stuck on a problem, ask yourself, 'What
> piece of information have I not used?'

> **Concept:** In many problems, there's more than meets the eye. Ex-
> tending segments that seem to end abruptly (particularly
> in the middle of a triangle) sometimes gives useful infor-
> mation.

Problem 5.17:   Flagpole $\overline{CD}$ is 12 feet tall. Flagpole $\overline{AB}$ is 9 feet tall. Both flagpoles are perpendicular to the ground. A straight wire is attached from B to D, and another from A to C. The flagpoles are 40 feet apart, and the wires cross at E, which is directly above point F on the ground. Find EF.

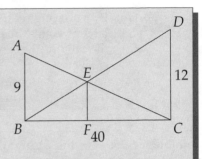

Solution for Problem 5.17: We start off by noticing that $\overline{AB} \parallel \overline{EF} \parallel \overline{CD}$ since all three are perpendicular to $\overline{BC}$. By now you know the drill: parallel lines mean similar triangles. We look first for similar triangles that include $\overline{EF}$, and we see $\triangle CEF \sim \triangle CAB$ and $\triangle EBF \sim \triangle DBC$. Therefore, we have

$$\frac{EF}{AB} = \frac{CF}{CB} = \frac{EC}{AC} \quad \text{and} \quad \frac{EF}{CD} = \frac{BF}{CB} = \frac{BE}{BD}.$$

We see CB in both groups, so we investigate the ratios involving CB more closely. We see that we have $CF + BF = CB$, so

$$\frac{EF}{AB} + \frac{EF}{CD} = \frac{CF}{CB} + \frac{BF}{CB} = \frac{CF + BF}{CB} = \frac{CB}{CB} = 1.$$

Now we can find EF:

$$EF = \frac{1}{\frac{1}{AB} + \frac{1}{CD}} = \frac{1}{\frac{1}{9} + \frac{1}{12}} = \frac{36}{7}.$$

Notice that the length of $\overline{BC}$ is irrelevant! □

WARNING!!   In the last solution we didn't spell out exactly why
☢ $\triangle EBF \sim \triangle DBC$, since we've gone through those steps several times already. When you are writing solutions for your class or for a contest, you should include the steps we left out here (cite which angles are equal and why, then invoke AA). Only start leaving out the simple steps if you are certain that it is O.K. to do so.

We finish this section by exploring why AA Similarity works.

Problem 5.18: In the diagram we have two triangles ($\triangle ABE$ and $\triangle ACD$) with equal angles, and sides with lengths as marked. Find BE and DE without using AA Similarity. Can you use your method to prove why AA Similarity works?

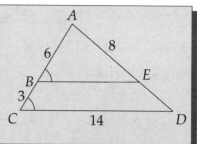

Solution for Problem 5.18: We would like to show that $\triangle ABE \sim \triangle ACD$, so we start thinking about side length ratios. The only ratio tool we have that doesn't depend on having similar triangles already is the Same Base/Same Altitude technique, so we try that. We don't have any triangles to use our technique on,

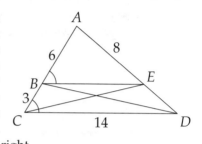

so we draw $\overline{BD}$ and $\overline{CE}$ as shown to the right.

$\triangle ABE$ and $\triangle AEC$ share an altitude from E, so

$$\frac{[ABE]}{[ACE]} = \frac{AB}{AC} = \frac{6}{9} = \frac{2}{3}. \tag{5.1}$$

Similarly, $\triangle ABE$ and $\triangle ABD$ share an altitude from B, so

$$\frac{[ABE]}{[ABD]} = \frac{AE}{AD} = \frac{8}{8 + DE}. \tag{5.2}$$

We suspect that $AB/AC = AE/AD$ because we suspect $\triangle ABE \sim \triangle ACD$. From (5.1) and (5.2), we have

$$\frac{AE}{AD} = \frac{[ABE]}{[ABD]} \quad \text{and} \quad \frac{AB}{AC} = \frac{[ABE]}{[ACE]}.$$

Since the numerators in our area ratios are the same, we need only show that $[ABD] = [ACE]$. These two areas share $[ABE]$, so we need only show that $[BEC] = [BED]$.

Since $\angle ABE = \angle ACD$, we know $\overline{BE} \parallel \overline{CD}$. Therefore, the altitudes from C and D to $\overleftrightarrow{BE}$ must be the same. Hence, triangles BEC and BED have the same base ($\overline{BE}$) and the same length altitudes to that base, so $[BEC] = [BED]$.

Finally, we can find DE. We have:

$$[AEC] = [AEB] + [BEC] = [AEB] + [BED] = [ABD],$$

so we can use our area ratios above. Since $[ABE]/[ACE] = [ABE]/[ABD]$, we have $AB/AC = AE/AD$, so

$$2/3 = 8/(8 + DE).$$

Solving this equation for DE, we find that $DE = 4$.

Now we very strongly suspect that $BE/CD = AE/AD$. To prove it, we use the same process we just followed.

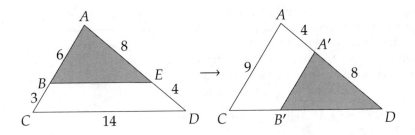

We consider $\triangle A'B'D$ where A' is on $\overline{AD}$ and B' is on $\overline{CD}$ such that $A'D = AE$ and $BE = B'D$. Since $\overline{BE} \parallel \overline{CD}$, we have $\angle AEB = \angle ADC = \angle A'DB'$. Therefore, $\triangle A'B'D \cong \triangle ABE$ by SAS Congruence. (You can also think of $\triangle A'DB'$ as the result of sliding $\triangle ABE$ along $\overline{AD}$ until side $\overline{BE}$ is on $\overline{CD}$.) Since $A'D = AE$, we have $AA' = ED = 4$. We also have $\angle DA'B' = \angle EAB = \angle DAC$, so $\overline{AC} \parallel \overline{A'B'}$.

We can chase areas around as before to show that $B'D/CD = A'D/AD$, so $B'D = (2/3)(14) = 28/3$. Since $B'D = BE$, we have $BE = 28/3$. Note that because $B'D = BE$ and $A'D = AE$, we have shown that $BE/CD = AE/AD$, as suspected. $\square$

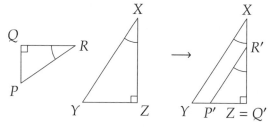

$\triangle PQR$ and $\triangle YZX$ in the diagram to the left have two angle measures in common (and consequently the third angles are equal, too). We can therefore move $\triangle PQR$ on top of $\triangle YZX$ such that two of the sides of the 'moved' triangle coincide with sides of $\triangle YZX$, as $\triangle P'Q'R'$ in the diagram shows.

We can use the exact same approach as we used in Problem 5.18 to show that if two angles of one triangle equal the corresponding angles of the other, then each pair of corresponding lengths in the two triangles has the same ratio.

> **Concept:** Area can be a very useful problem solving tool even in problems that appear to have nothing to do with area.

Exercises for Section 5.5

5.5.1 X and Y are on sides $\overline{PQ}$ and $\overline{PR}$, respectively, of $\triangle PQR$ such that $\overline{XY} \parallel \overline{QR}$. Given $XY = 5$, $QR = 15$, and $YR = 8$, find PY.

5.5.2   In the figure, the area of $\triangle EDC$ is 25 times the area of $\triangle BFD$.

(a)   Find CD/DB.

(b)   Find $[EDC]/[ABC]$.

(c)★   Find $[AFE]/[ABC]$.

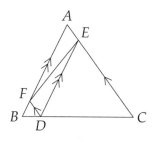

5.5.3   In the diagram, $\overline{WZ} \parallel \overline{XY}$ and $\overline{WX} \parallel \overline{ZY}$. $\overline{WA}$ and $\overline{WB}$ hit $\overline{XZ}$ at C and D, respectively, such that $ZC = XD$.

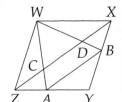

(a)   Prove that $ZC/XC = AC/WC$.

(b)   Prove that $XD/ZD = DB/WD$.

(c)   Prove $\overline{CD} \parallel \overline{AB}$.

5.5.4   In the diagram at left below, $PQ = PR$, $\overline{ZX} \parallel \overline{QY}$, $\overline{QY} \perp \overline{PR}$, and $\overline{PQ}$ is extended to W such that $\overline{WZ} \perp \overline{PW}$.

(a)   Show that $\triangle QWZ \sim \triangle RXZ$.

(b)★   Show that $YQ = ZX - ZW$.

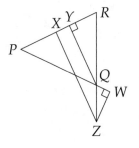
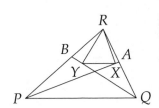

5.5.5★ $\overline{PA}$ and $\overline{BQ}$ bisect angles $\angle RPQ$ and $\angle RQP$, respectively. Given that $\overline{RX} \perp \overline{PA}$ and $\overline{RY} \perp \overline{BQ}$, prove that $\overline{XY} \parallel \overline{PQ}$.

5.6 Summary

> **Definition:** Two figures are **similar** if one is simply a blown-up, and possibly rotated and/or flipped, version of the other.
>
> > **Important:** Corresponding angles in similar figures are equal, and the ratio of the lengths of corresponding sides of similar triangles is always the same.
> >
> >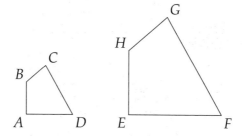
> >
> > In similar quadrilaterals $ABCD$ and $EHGF$, we have $\angle A = \angle E$, $\angle B = \angle H$, $\angle C = \angle G$, and $\angle D = \angle F$. We also have
> >
> > $$\frac{AB}{EH} = \frac{BC}{HG} = \frac{CD}{GF} = \frac{DA}{FE} = \frac{AC}{EG} = \frac{BD}{HF}.$$
> >
> > We denote these figures as similar by writing
> > $$ABCD \sim EHGF.$$

There are three main ways to show that two triangles are similar:

- **AA Similarity.** If two angles of one triangle equal two angles of another, then the triangles are similar. This is by far the most commonly used method to prove two triangles are similar. (Section 5.2)

- **SAS Similarity.** If two sides in one triangle are in the same ratio as two sides in another triangle, and the angles between the sides in each triangle equal each other, then the triangles are similar. (Section 5.3)

- **SSS Similarity.** If each side of one triangle is the same constant multiple of the corresponding side of another triangle, then the triangles are similar. (Section 5.4)

Parallel lines and perpendicular lines are clues to look for similar triangles. Three very common set-ups that contain similar triangles are shown below.

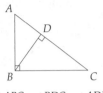

$\triangle ABC \sim \triangle BDC \sim \triangle ADB$

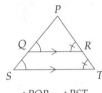

$\triangle PQR \sim \triangle PST$

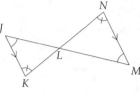

$\triangle JLK \sim \triangle MLN$

> **Important:** If $\overline{BC} \parallel \overline{DE}$ and $\overleftrightarrow{BD}$ and $\overleftrightarrow{CE}$ meet at A as shown, then
>
> $$\frac{AB}{BD} = \frac{AC}{CE}.$$

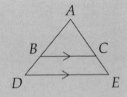

> **Important:** If two triangles are similar such that the sides of the larger triangle are k times the sides of the smaller, then the area of the larger triangle is k^2 times that of the smaller.
>
> This relationship holds for any pair of similar figures, not just for triangles.

Problem Solving Strategies

> **Concepts:**
>
> - When you're stuck on a problem, ask yourself, 'What piece of information have I not used?'
>
> - In many problems, there's more than meets the eye. Extending segments that seem to end abruptly (particularly in the middle of a triangle) can often yield quick solutions.
>
> *Continued on the next page. . .*

Concepts: . . . *continued from the previous page*

- When stuck on a problem, try solving an easier related problem. For constructions, useful easier related problems often involve relaxing one of the constraints of the problem.

- Consider using similar triangles in problems involving ratios of segment lengths.

Things To Watch Out For!

WARNING!! Below are shown two common situations that lead to mistakes. The diagram on the left may lead you to write '$\triangle ABC \sim \triangle ADE$, so $AB/BD = BC/DE$.' The one on the right might lead to '$\triangle JKL \sim \triangle NLM$, so $JL/NL = KL/ML$.' Both of these are **incorrect**! Make sure you see why!

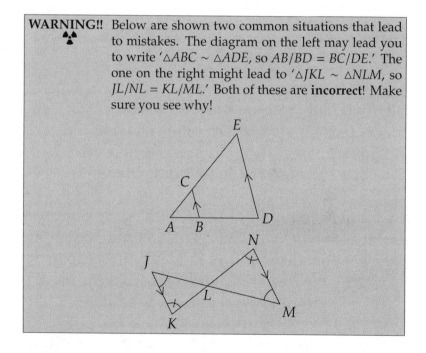

Review Problems for Chapter 5

Many Review and Challenge problems for this chapter are present in the textbook *Introduction to Geometry*.

WORLDWIDE ONLINE OLYMPIAD TRAINING

WOOT!

The Worldwide Online Olympiad Training program is a 7-month preparation and testing program that brings together many of the best students from around the world to learn Olympiad problem solving skills. The program includes:

- 16 problem solving classes.
- 8 problem solving articles and sets of practice problems.
- 2 practice AIME-style tests.
- 3 practice Olympiad tests.
- 3 problem sets.
- Detailed feedback on the Olympiad tests and problem sets.
- Dedicated message board.
- 24-hour online classroom access.

More than half the 2006 USA Mathematical Olympiad Winners participated in 2005-06 WOOT, as did several students who qualified to represent their countries in the 2006 International Math Olympiad.

Due to sponsorship from D.E. Shaw group and Art of Problem Solving, all 2006 Math Olympiad Summer Program participants are invited to participate in WOOT for free. This offers students the most outstanding peer group possible.

Visit www.artofproblemsolving.com for more information about WOOT!

"I just [want to] thank all my WOOT teachers and classmates for helping me to get to my first IMO [International Mathematical Olympiad]...I'm very glad to have been a part of this and I can really say that the classes that I took were great and helped a lot while I was taking my tests."

6

What is Problem Solving?

Note: This essay was written by Richard Rusczyk. This essay, and many other similar essays, can be found by going to www.artofproblemsolving.com and clicking on "Resources".

I was invited to the Math Olympiad Summer Program (MOP) in the 10th grade. I went to MOP certain that I must really be good at math. In my five weeks at MOP, I encountered over sixty problems on various tests. I didn't solve a single one. That's right - I was 0-for-60+. I came away no longer confident that I was good at math. I assumed that most of the other kids did better at MOP because they knew more tricks than I did. My formula sheets were pretty thorough, but perhaps they were missing something. By the end of MOP, I had learned a somewhat unsettling truth. The others knew fewer tricks than I did, not more. They didn't even have formula sheets!

At another contest later that summer, a younger student, Alex, from another school asked me for my formula sheets. In my local and state circles, students' formula sheets were the source of knowledge, the source of power that fueled the top students and the top schools. They were studied, memorized, revered. But most of all, they were not shared. But when Alex asked for my formula sheets I remembered my experience at MOP and I realized that *formula sheets are not really math.* Memorizing formulas is no more mathematics than memorizing dates is history or memorizing spelling words is literature. I gave him the formula sheets. (Alex must later have learned also that the formula sheets were fool's gold - he became a Rhodes scholar.)

The difference between MOP and many of these state and local contests

I participated in was the difference between problem solving and what many people call mathematics. For these people, math is a series of tricks to use on a series of specific problems. Trick A is for Problem A, Trick B for Problem B, and so on. In this vein, school can become a routine of "learn tricks for a week - use tricks on a test - forget most tricks quickly." The tricks get forgotten quickly primarily because there are so many of them, and also because the students don't see how these "tricks" are just extensions of a few basic principles.

I had painfully learned at MOP that *true mathematics is not a process of memorizing formulas and applying them to problems* tailor-made for those formulas. Instead, the successful mathematician possesses fewer tools, but knows how to apply them to a much broader range of problems. We use the term "problem solving" to distinguish this approach to mathematics from the "memorize–use–forget" approach.

After MOP I relearned math throughout high school. I was unaware that I was learning much more. When I got to Princeton I enrolled in organic chemistry. There were over 200 students in the course, and we quickly separated into two groups. One group understood that all we would be taught could largely be derived from a very small number of basic principles. We loved the class - it was a year long exploration of where these fundamental concepts could take us. The other, much larger, group saw each new destination not as the result of a path from the building blocks, but as yet another place whose coordinates had to be memorized if ever they were to visit again. Almost to a student, the difference between those in the happy group and those in the struggling group was how they learned mathematics. The class seemingly involved no math at all, but those who took a memorization approach to math were doomed to do it again in chemistry. The skills the problem solvers developed in math transferred, and these students flourished.

We use math to teach problem solving because it is the most fundamental logical discipline. Not only is it the foundation upon which sciences are built, it is the clearest way to learn and understand how to develop a rigorous logical argument. There are no loopholes, there are no half-truths. The language of mathematics is precise, as is "right" and "wrong" (or "proven" and "unproven"). Success and failure are immediate and indisputable; there isn't room for subjectivity. This is not to say that those who cannot do math cannot solve problems. There are many paths to strong problem solving skills. *Mathematics is the shortest.*

Problem solving is crucial in mathematics education because it transcends mathematics. By developing problem solving skills, we learn not only how to tackle math problems, but also how to logically work our way through any problems we may face. The memorizer can only solve problems he has encountered already, but the problem solver can solve problems she's never seen before. The problem solver is flexible; she can diversify. Above all, she can *create*.

FREE ONLINE RESOURCES FOR PROBLEM SOLVING MATHEMATICS AT WWW.ARTOFPROBLEMSOLVING.COM!

The Art of Problem Solving website includes many free resources for teachers and students of problem solving mathematics.

- **FORUM:** Our forum currently has over 22,000 members and over half a million posts. It has message boards for middle and high school students, including forums for prominent math contests like MATHCOUNTS and the AMC. Membership in the community is FREE!
- **AOPS WIKI:** An online project supporting educational content useful to students and teachers of math, science, computer science, technology, and other topics. Content is added, edited, and updated by members.
- **ARTICLES:** Articles about preparation for events, building a math team, math topics, and more. Written by Art of Problem Solving staff, successful math coaches, professors, and students.
- **LINKS:** Links to numerous resources such as local, regional, and national events and competitions, summer programs, scholarships, other websites, and more!
- **BOOKS:** Important preparatory and motivational books of interest to students and teachers.
- **LaTeX GUIDE:** A guide for students to learn how to typeset mathematics.

VISIT WWW.ARTOFPROBLEMSOLVING.COM FOR MORE INFORMATION!